Numanities - Arts and Humanities in Progress

Volume 13

Series Editor

Dario Martinelli, Kaunas University of Technology, Kaunas, Lithuania

The series originates from the need to create a more proactive platform in the form of monographs and edited volumes in thematic collections, to discuss the current crisis of the humanities and its possible solutions, in a spirit that should be both critical and self-critical.

"Numanities" (New Humanities) aim to unify the various approaches and potentials of the humanities in the context, dynamics and problems of current societies, and in the attempt to overcome the crisis.

The series is intended to target an academic audience interested in the following areas:

- Traditional fields of humanities whose research paths are focused on issues of current concern;
- New fields of humanities emerged to meet the demands of societal changes;
- Multi/Inter/Cross/Transdisciplinary dialogues between humanities and social and/or natural sciences;
- Humanities "in disguise", that is, those fields (currently belonging to other spheres), that remain rooted in a humanistic vision of the world;
- Forms of investigations and reflections, in which the humanities monitor and critically assess their scientific status and social condition;
- Forms of research animated by creative and innovative humanities-based approaches;
- Applied humanities.

More information about this series at http://www.springer.com/series/14105

Giedre Valunaite Oleskeviciene ·
Jolita Sliogeriene

Social Media Use
in University Studies

 Springer

Giedre Valunaite Oleskeviciene
Institute of Humanities
Mykolas Romeris University
Vilnius, Lithuania

Jolita Sliogeriene
Department of Foreign Languages
Vilnius Gediminas Technical University
Vilnius, Lithuania

ISSN 2510-442X   ISSN 2510-4438 (electronic)
Numanities - Arts and Humanities in Progress
ISBN 978-3-030-37726-7   ISBN 978-3-030-37727-4   (eBook)
https://doi.org/10.1007/978-3-030-37727-4

This Springer imprint is published by the registered company Springer Nature Switzerland AG
The registered company address is: Gewerbestrasse 11, 6330 Cham, Switzerland

Acknowledgements

We are grateful to Mykolas Romeris University, the Institute of Humanities, and Vilnius Gediminas Technical University, the Faculty of Creative Industries for the environment conducive to research that inspired and equipped us with the means necessary for the research.

We are also thankful to our colleague and main advisor Prof. Dr. Dario Martinelli who has given generous support, encouragement, and numerous valuable insights throughout the process of monograph preparation.

Many thanks go to our international project partners within the framework of the Grundtvig multilateral partnership project "Institutional Strategies Targeting the Uptake of Social Networking in Adult Education (ISTUS)". An international inductive qualitative study served as a starting point of the research that captured the initial insights on social media use in university studies.

Our special gratitude goes to our insightful reviewers who had to remain anonymous but whose advice was of immeasurable value.

Special thanks to our family members who have been patient sufferers of lacking our proper attention.

Introduction

The emergence of Web 2.0 technologies and their applications such as social media continues to impact educational processes throughout the world. Some theoreticians and researchers explore the ways in which mastering new social technologies are able to ensure better quality of life. Others warn against the risk of mixing reality with illusory concepts and the appearance of simulated reality, which can result in manipulative projects that distort natural human life. Despite a range of opinions, all commentators agree that the role of social media is increasing and changing the ways in which we accept and process information.

According to recent research, approximately 95% of young people (age groups 13–17 and 18–29) use social media regularly (Lenhart et al. 2010). The study by Sutherland et al. (2018) shows that more than half of students (52.8%) identify that university social media profiles help them to feel part of the academic community. Although the study by Akakandelwa and Walubita (2017) reveals some negative impact of social media, the authors admit that social interaction is enhanced by social media use among students and recommend productive use of social media in order to minimize its negative impact. Applications of Web 2.0 technologies are used more and more especially by young people in widely varying places and social contexts (Hargittai 2007). Researchers have identified that a generation of future students are active users of social media, especially in creating digital content for use in digital media.

Web 2.0 technologies change social distributions and have the potential to create new possibilities and reshape our links to objects, places, and each other, yet this has not been sufficiently explored (Beer and Burrows 2007). Research into the impact of social media on pedagogics and social links in education is a growing field of research (Selwyn 2007). In this regard, current research emphasizes student use of social media for the purpose of study (Ellison, Steinfield, and Lampe 2007; Selwyn 2007). It is also focused on educators' use of social media in their teaching practices (Hewitt and Forte 2006; Mason and Rennie 2008; Mazer, Murphy, and Simonds 2007), transformation of teaching practices while using social media (Cheal, Coughlin, and Motore 2012) and pedagogue professional transformation (Oberg and Bell 2012). Research on the application and influence of social media on university studies provide an extensive picture of technology and social media influence on the development of university studies and their democratization processes (Bach, Haynes and Smith 2007). The research also considers the development of British universities and is inspired by both political decisions and technology application (Cowen 2013).

In the context of Lithuania, research on the use of social media in education is rapidly acquiring national significance. Petkūnas and Jucevičienė (2006) analyze the change of educational paradigm due to the influence of information technology application. Zygmantas (2007) focuses on the changing requirements for pedagogues. Assessment of technology-based education content has been recently discussed in the national context (Volungevičienė and Tereseviečienė 2011). Duoblienė (2011) provides an extensive introspection into the development and issues of the Lithuanian educational system in the broader European context, presenting insightful remarks concerning technology and social media application in education. The development of the idea of university and university studies, as well as university democratization processes, is also analyzed (Kraujutytė 2002; Samalavičius 2010).

In 2014, Prensky introduced the term "VUCA" (variability, uncertainty, complexity and ambiguity), which stresses the growing complexity of our learning and living environments. The term "VUCA" embraces the variability of the education technologies—what seems appropriate today may not be chosen tomorrow—uncertainty of life paths, the disappearance of reliable "given" choices that lead to stable futures, the complexity of the educational and developmental environments—which embrace ever-changing and growing technologies and the world itself—and, finally, pervasive ambiguity, when our worst students by some measures are the best by other measures. Educational problems tend to persist, as the variability of education technologies seems to pose a challenge to the teaching staff or university teachers (further teachers) who may still be grappling with new emerging technologies of social media.

VUCA expressly defines our ever-changing education environments. Teachers cannot shut off the realities of the modern world, which bring social media into the university study process; they have to accept the reality of social media. Similarly, the problem is described by McLuhan (2003), who indicates the demise of the era of mechanistic linear philosophy prevalence. The author states that linearity has been replaced by the simultaneity and concentric nature of the digital age, with the endless intersection of projections where all types of social media constantly interact with each other. Thus, all participants in the university study process, i.e., teaching staff, students, and administrators, have to move into the simultaneous worlds of social media used in the study process.

Selwyn (2012) observes that the research field of social media application in higher education embraces multiple discourses ranging from absolutely enthusiastic ones featuring social media as panacea to the most resistant ones viewing social media as a totally disruptive technology contaminating education and human minds. The research field is comparatively new and developing, still embracing many unanswered questions and featuring the prevailing tendencies to apply a constructivist approach in looking at how to enhance social media use in university studies for teaching and learning purposes and at how to identify procedures that can be easily organized, applied, and evaluated. The question of the human factor, however, seems to be marginalized. Attentiveness, the pedagogical relationship, and the human being in the study environments saturated with social media technologies cannot be easily counted, but they are also important in education. In this context, research on the phenomenon of social media application in university studies is absolutely relevant and new as it aims to look deeper into the assumptions about the "almighty" properties of social media in education and also to find out if social media could have some enhancing effect on educational processes.

Research object. The research object is the meaning of social media use in university studies. The research investigates the phenomenon of social media use in university studies with a particular focus on the meaning of the "lived experience" of the university student participants.

Research aim and objectives. This investigation is an inductive qualitative research with a phenomenological approach; it contributes to the broad research field with multiple approaches to the use of social media in university studies. The aim of the present study is to investigate how university study participants: teachers, students, and administrators, make sense of social media use in university studies through their own lived experiences. The meaning is revealed through the exploration of teacher, student, and administrator personal stories of social media use in university studies. Pursuing the research aim, the following research objectives have been set:

1. To overview the discourse of social media use in university studies;
2. To provide insights on social media use in university studies through a phenomenological approach;
3. To identify how research participants—teachers, students, and administrators—make meaning of their experiences of using social media in university studies;
4. To disclose and structuralize the multifaceted nature of the phenomenon of social media use in university studies.

Research significance. The multiple discourses on social media use in university studies are often contradictory and include various views ranging from enthusiastic ones to critical perceptions of social media as disruptive technology. The research on the phenomenon of social media use in university studies is thus a scientific research input into the vast field of the research on social media's educational use. Educational context is an important factor for learning, as has been proved by educationalists (Lave and Wenger 2002), so the contextual realities are important while researching social media technology use in university studies. The research will create better understanding of social media use in university studies by revealing the meaning of "human being" in university study environments enriched by social media. The results of the research will enable recommendations for social media use in university studies and also reveal areas for future research.

Methodology of the research (methods and implementation). Inductive qualitative research is sensitive to human activity in educational environments and admits that humans, based on their experiences, continuously and actively create their own unique educational realities and can successfully complement quantitative research paradigm (Denscombe 2003; Silverman 2005; Thomas 2006; Creswell 2007; Elo and Kyngäs 2007; Mayring 2014; Biesta 2013). A phenomenological perspective acknowledges the inevitable relative subjectivity and research relativity and also recognizes the value of obtaining narratives about the topic (Heidegger 1972; Gadamer 1999; Stiegler 2010; Saevi 2012; Adams 2012; van Manen 2014; Smith 2009). Before carrying out the research, a literature overview is conducted, aiming to reveal the complexity of the notion of "social media" and the research field of social media use in the study process as well as to present the multiple discourses on social media use in university studies.

The starting point of the study is international inductive qualitative research, which aims to gain initial insights into social media use in university studies. It was carried out within the framework of the Grundtvig multilateral partnership project "Institutional Strategies Targeting the Uptake of Social Networking in Adult Education (ISTUS)" (De Angelis et al. 2013). The research is based on problem-centered interviews, the main characteristic of which is the prompting by questions and statements and having a chance to ask ad hoc questions (Witzel and Reiter 2012). The study process is shaped by educators—teaching staff and administrators—although the views and attitudes of learners are important too, especially bearing in mind the informal character of social media; thus, three groups of research participants: teachers, students, and administrators, are included in the interview series to ensure data source triangulation.

The starting point leads to in-depth qualitative inductive research with a phenomenological approach in home institutions. In-depth interviews with teachers, students, and administrators are carried out at home institutions; they aimed to explore social media use in university studies at a deeper level. The research is focused on how teachers, students, and administrators make sense of their experiences related to social media use in studies. Inductive qualitative analysis is enriched by the phenomenological approach, which is perceived as not pure description but as an interpretation of lived experiences related to social media use in university studies. The existentials of lived experience indicated by van Manen (2014): relationality, spatiality, and temporality, are used while interpreting interview data.

While analyzing the data, the collected empirical material is overviewed, and the meaningful statements, sentences, or other extracts, which provide the information related to the participant experience connected to social media use in university studies, are highlighted. Thomas (2006), Silverman (2005), and Elo and Kyngäs (2007) refer to this process of analysis as reading the data "horizontally," grouping the segments of text into emerging themes. Then, the themes forming clusters of meaning are linked into super-ordinate themes. Descriptions and quotations to illustrate the meanings of the themes are presented.

The structure of the monograph. The monograph is organized into four chapters. Chapter 1 is introductory in nature and sets the context of the research. It contains a literature review that provides a definition of "social media" and analyzes social media use in university studies, emerging changes in university studies and emerging educational approaches to integrating social media into university studies. Part of Chap. 1 is also devoted to practicing insights on technology and education, including social media use in university studies while applying a phenomenological approach. It embraces the philosophical background and phenomenological insights on education and technology.

Chapter 2 is devoted to the research methodology. It explores the qualitative inductive research with its phenomenological approach and provides the methodology used in the research. Chapter 3 reports the research findings of the international qualitative inductive research with a phenomenological approach as the starting point of the research and the research findings of the in-depth qualitative inductive research with a phenomenological approach in home institutions. The multifaceted nature of the phenomenon of social media application in the university study process is extensively discussed. Conclusion brings the monograph to a close.

Concepts Used in the Monograph

Communication is a process by which information is exchanged between individuals through a common system of symbols, signs, or behavior (Merriam-Webster Online Dictionary).

Communications are means of sending messages, orders, etc., including telephone, telegraph, radio, and television (Merriam-Webster Online Dictionary).

Creativity is the tendency to generate or recognize ideas, alternatives, or possibilities that may be useful in solving problems, communicating with others, and entertaining ourselves and others (Franken 1994). From the perspective of systems theory on creativity, there are three interacting systems: the individual, the social environment, and culture, which define creativity (Csikszentmihalyi 1996).

Education could be defined as the process of activating learning or the acquisition of knowledge, skills, values, beliefs, and habits. Education is more than fostering understanding and an appreciation of emotions and feelings. It is also concerned with change—"with how people can act with understanding and sensitivity to improve their lives and those of others" (Smith and Smith 2008).

Higher education is education beyond the secondary level, *especially*: education provided by a college or university. Institutions of higher education include not only colleges and universities but also professional schools in fields such as law, theology, medicine, business, music, and art. They also include teacher-training schools, community colleges, and institutes of technology. At the end of a prescribed course of study, a degree, diploma, or certificate is awarded (Kraujutytė, 2002).

Inductive qualitative research is often referred to as a "bottom-up" approach to knowing, in which the researcher uses observations to build an abstraction or to describe a picture of the phenomenon that is being studied. The inductive approach enables researchers to identify key themes in the area of interest by reducing the material to a set of themes or categories (Lodico, Spaulding, and Voegtle 2010).

Lived experience as a concept (from the German *Erlebnis*) possesses special methodological significance. The notion of "lived experience," as used at the end of the nineteenth and at the beginning of the twentieth century in the works of Dilthey (1985), Husserl (1970), and their contemporary exponents, announces the intent to explore directly the origins or prereflective dimensions of human existence. The etymology of the English term "experience" does not include the meaning of lived—it derives from the Latin *experientia*, meaning "trial, proof, experiment, experience"—but the German word for experience, *Erlebnis*, already contains the word *Leben*, meaning "life" or "to live". The verb *erleben* literally means "living through something". At the end of the nineteenth century, Wilhelm Dilthey (1985) offered the first systematic explication of lived experience and its relevance for the human sciences. He describes "lived experience" as a reflexive or self-given awareness that inheres in the temporality of consciousness of life as we live it (van Manen 2014).

Pedagogical (educational) relationship designates a special kind of personal relationship between adult and child or teacher and student that is different from other personal relationships. The pedagogical relation is described in the humanistic European pedagogical tradition. It has also been discussed more recently by Max van Manen (van Manen 2014). Through an educational relationship, the teacher presents the learner not merely with some knowledge already possessed but also with the condition for recognizing it as truth. A student is open to the gift of teaching; a student does not limit himself/herself to the task of learning but is open to the possibility of being taught—be it school education or another form of education as an adult—it is never that students just learn, as learning implies something that one can only do for oneself. The role of the teacher is to re-instigate dialogical and reflective practices that in turn link people back to the world. Reflecting the importance of being taught, the educational relationship stands out as having key importance (Biesta 2013).

Pedagogy is most commonly defined as the approach to teaching or the art and science of teaching. It refers broadly to the theory and practice of teaching, and how this influences the growth of learners (Scardamalia and Bereiter 2006).

Phenomenological existentials are helpful universal "themes" with which to explore meaning aspects of our life world and of the particular phenomena that we may be studying; for example, the notions of lived relation, body, space, time, and things are existentials in the sense that they belong to everyone's life world—they are universal themes of life (van Manen 2014).

Phenomenology (from the Greek *phainómenon*, meaning "that which appears," and *logos*, meaning "study") is the philosophical study of the structures of experience and consciousness. As a philosophical movement, it was founded in the early years of the twentieth century by Edmund Husserl and was later expanded upon by a circle of his followers at the universities of Gottingen and Munich in Germany. It then spread to France, the USA, and elsewhere, often in contexts far removed from Husserl's early work (Zahavi 2003).

Phenomenon—*The Oxford English Dictionary* presents the following definition: something (such as an interesting fact or event) that can be observed and studied and that typically is unusual or difficult to understand or explain fully. The Greek *phainómenon* means "appearance". Phenomena are what occur in the mind: Mental phenomena are acts of consciousness (or their contents), and physical phenomena are objects of external perception. Phenomena are whatever we are conscious of: objects and events around us, other people, ourselves—even (in reflection) our own conscious experiences as we experience them (van Manen 2014).

University studies include studies at an institution of higher education offering tuition in mainly non-vocational subjects and typically having the power to confer degrees. University studies provide students with integrated, connected learning experiences that lay the foundation for lifelong intellectual development (Jucevičienė et al. 2010).

Contents

Chapter 1
Social Media in University Studies: An Overview

1.1 Theoretical Foundations of the Research

The chapter presents the ongoing debate on social media use in university studies and the issue of interplay between new technologies and pedagogies. It also introduces the phenomenological philosophical background giving the basis for the phenomenological approach. Phenomenological insights on human–technology relations are discussed as well.

1.1.1 Complexity of Social Media Educational Use in University Studies

Many researchers acknowledge the prevalence of social media, its availability, and its potential to be used in learning environments promoting active learning in university studies (Anderson 2007; Eijkman 2008; Lee and McLoughlin 2010; Selwyn 2012). Selwyn (2012) states that there are three main reasons for using social media in university studies or higher education:

- Contemporary students start using social media abundantly long before entering the institutions of higher education, which makes them used to the connected state, sharing and working collectively;
- The changing relationship of university students to knowledge creation and the settings of formal education;
- The emergence of user-driven education.

G. Valunaite Oleskeviciene and J. Sliogeriene, *Social Media Use in University Studies*, Numanities - Arts and Humanities in Progress 13, https://doi.org/10.1007/978-3-030-37727-4_1

There are, however, opposite opinions; for example, Friesen and Lowe (2012) assert that social media was not originally constructed for formal education, and they point out that there is a lack of developing skills such as debate and expressing opposing views, which are considered necessary in learning environments.

Prensky (2014) discusses the increasing complexity of the environments where people learn and live by introducing the term VUCA which contains variability of the education technologies, the author observes that what seems appropriate today may not be chosen tomorrow, there is also uncertainty of life paths because reliable "given" choices which lead to stable future seem to have been disappearing, complexity of the educational and developmental environments also seems to be increasing, because such environments embrace ever-changing and growing technologies and the world itself, and finally pervasive ambiguity is characteristic of learning environments because at times our worst students by some measures are the best by some other measures.

Moore (1993) indicates that online interaction is transactional in distance and time. The "transactional nature of time" element is evident in the contexts of the durations of activities where time frames or limits are important in online discussions (Jeong and Frassier 2008). Jeong and Frassier (2008) investigated how time limits may affect online discussions.

The use of online technologies is turning out to become an important challenge for academic staff as e-learning has impacted and continues to affect higher education in global and local contexts (Donnelly 2014). There is a growing enthusiasm among academic staff and as well some pressure from students for the lecturers to use new technologies, even if explicit institutional policies are lacking. The characteristics of blended learning in the literature are promising increased learning and student engagement, as well as collaboration. The issue of interplay between new technologies and pedagogies remains the question to be researched. Gredler (2005) suggests that the role of technology in teaching and learning remains an issue for theory development and research as there is a qualitative difference between "teaching online" and "putting a course online."

Bates (2005) advocates that the lessons learned while applying technology in education should be kept in mind while applying new emerging technologies; however, as the author observes, the past lessons with the technology application, such as the need to redesign and reorganize teaching in order to successfully and fully apply the new technology, are often ignored. In addition, what should be kept in mind is that technologies do not simply roll in—there always have to be efforts to address certain groups of people, making sure they get access to the technology.

Selwyn (2012) analyzes the place of social media in higher education concerning three main lines: the learner, learning, and higher education provision. The author supports the insights of other scholars (Papacharissi 2010; Subrahmanyam and Šmahel 2011), claiming that as new cohorts of students enter higher education the students who are used to digital juggling of their activities and increased autonomy of social activity, being able to choose what they do, when, where, and how, it becomes important for universities to connect with these students. It is identified, however, that traditional top-down institutions of higher education are poorly equipped to meaningfully engage their students, as the author observes: "Even the best-intentioned universities are able only to offer their students an artificially regulated and constrained engagement with social media" (Selwyn 2012). There is a certain clash between a hierarchically structured way of communication and learning offered by universities as institutions on the one hand and the linear ways of social media on the other. It is also observed by Ulbrich et al. (2011) that:

> Members of the net generation use the web differently, they network differently, and they learn differently. When they start at university, traditional values on how to develop knowledge collide with their values. Many of the teaching techniques that have worked for decades do not work anymore because new students learn differently too. The net generation is used to networking; its members work collaboratively, they execute several tasks simultaneously, and they use the web to acquire knowledge. (Ulbrich et al. 2011, p. 241)

Speaking about learning, Thomas and Seely-Brown (2011) state that learning in a social media context is based more on collective exploration and innovation, and individual instruction characteristic of formal education is less preferred. This goes in line with connectivism, the term coined by Siemens (2004), which stresses the ability to actively access information and augment it rather than passively retain information, which used to be a traditional way of teaching and learning in the environments of formal higher education. Many scholars admit that universities have the potential to use social media for collective knowledge creation; however, there are some critical attitudes like that of Carr (2010), who claims that students while using social media "are evolving from cultivators of personal knowledge into hunters and gatherers in the electronic data forest. In the process, we seem fated to sacrifice much of what makes our minds so interesting." As Selwyn (2012) observes, though, such critical attitudes are rarely based on extensive research.

Concerning formal higher education provision, a kind of tension remains between the scholars who believe that social media could be applied in higher education to its advantage and the ones who are skeptical and see social media as a disruptive technological tool that may unsettle the university in its existing form. Web 2.0, the term invented by DiNucci in 1999, that has stimulated the emergence of social media such as Twitter, Instagram, Facebook, blogs, and various applications allows users to interact and collaborate with each other through social media dialogue as creators of user-generated content in a virtual community. Many higher education researchers believe that universities are able to embrace social media and start discussing the emergence of "2.0 pedagogy," which contains innovative approaches to social media use (Lee and McLoughlin 2010). There are, however, also attitudes that social media poses a challenge to the very concept of traditional formal university education, and examples like Academic Earth, YouTube EDU, and the International University of the People are given to support the views that campus universities are becoming anachronistic in our digital technology era (Suoranta and Vadén 2010).

Selwyn (2012) observes that the ongoing debate is still not based on substantial research and is of a more speculative nature. In fact, social media use in higher education is not wholly positive or totally negative and should be analyzed in more disputable terms including advantages and disadvantages. The author also observes

that the wider context of social media use in higher education remains contradictory as well. First, there remains inequality in access to Internet and social media
tools, and the digital divide remains great depending on socioeconomic status, social
class, race, gender, geography, age, and educational background (Jones and Fox
2009). Also, democratic activity of social media appears to be questionable as clear
socioeconomic inequalities are observed in the contexts of social media, and social
media environments are not more socially integrated than the offline ones (Mayer
and Puller 2008). In addition, not all social media activities are related to educational contexts. Selwyn's (2009) study reveals that around 95% of college students'
interactions on social media were not related to their studies, so it seems rational
that Hosein et al. (2010) introduce terms like "living technologies," which are used
for everyday social interaction and learning technologies used in educational environments for study purposes. Another concern is that optimistic expectations about
social media-enhanced collective creativity seem to be far-fetched. The majority of
users of social media applications prefer passive use of knowledge; user-creative
activities are largely limited to profile creation. Such a situation could be characterized by the economical term "logic of collective action" when a majority uses the
content created by a minority (Selwyn 2012). In fact, the situation could be described
in Gouseti's (2010) words as "a cycle of hype, hope, and disappointment," reflecting
the situation when at the initial state there is sometimes exaggerated enthusiasm,
which later develops into sober understanding. In this more realistic context, Selwyn
(2012) identifies two major issues: the discussion on the nature of the institutions
of formal higher education, including debates about the nature of institutionalized
education, and integration of social media into educational environments.

1.1.2 Cultivating Insight: A Phenomenological Approach to Education and Technology

The impact of phenomenology as philosophical thought extends beyond philosophy
and is used in multiple fields of social sciences such as sociology, psychology, politics, and education. The interest in phenomenology probably stems from the wide
context of human studies and extensive efforts to understand complex phenomena
in the dynamic and mosaic postmodern world. The phenomenological philosophical
background is overviewed for providing the basis of the phenomenological approach.
Also, note is taken of phenomenological insights on human–technology relations and
a reshaping impact of the technology use in educational environments.

1.1.2.1 Philosophical Background

As for the philosophical foundations, according to German philosopher and mathematician Husserl (1859–1938) the phenomenological research focuses on what is

experienced in individual consciousness. It seeks to systematically examine individual consciousness or lived experience in order to return our experience to its roots—to abstract the essential structures, suspending subjectivity.

Heidegger raised the question of whether there exists objective knowledge beyond the limits of human perception (Denscombe 2003). He suggested that we cannot ever fully access these abstract structures because our observation is always colored by our subjectivity. The best we can do is to interpret. One of the key concepts from Heidegger's philosophy is "Dasein" (being-there, being-in-the-world). According to Heidegger, we are thrown into the world (Dasein), and so we cannot "objectively" observe the world because we are in a constant relationship with it. Man is always in the world; he defines himself in the process of life.

Merleau-Ponty thought in a similar way that we experience the world through our own bodies (Smith et al. 2009). We only know it through the body, so our experience is limited to our embodied nature. The body-subject gets the knowledge of the world through the body experience and through interpreting the experience. In this way, Husserl, Heidegger, and Merleau-Ponty are the main figures in phenomenological philosophy. Husserl's work reveals how important it is to focus on the experience and its perception. Heidegger and Merleau-Ponty contribute by treating a human being as immersed in the world of objects and relations, and in the world of language and culture. They tend toward the interpretive approach, which is based on the notion that a human being exists in the lived world, which is personally experienced by each separate individual. Phenomenology is interested in the subjective and conscious human being, emphasizing human wholeness, uniqueness, and individuality. Human-free will and responsibility are also emphasized.

Another important theoretical basis for interpretation in phenomenology derives from hermeneutics—interpretative theory (Smith et al. 2009). Again, Heidegger is a pioneer in hermeneutic phenomenology by merging together two philosophical thoughts. The beginning of hermeneutics attempts to interpret texts; however, Heidegger applied it in a broader philosophical context. Heidegger emphasized the person's background and "situatedness" in the world, stating that the background delineates the ways in which a person understands the world, and, through this understanding, one determines what is real and perceives the world. To paraphrase Munhall (1989), Heidegger had a view of people and the world as indissolubly related in cultural, in social, and in historical contexts. He described this relation as an indissoluble unity between a person and the world. Meanings appear as we are constructed by the world, while at the same time we are constructing the world from our own background and experiences. There is a constant transaction between the individual and the world as they constitute and are constituted by each other (Munhall 1989). Hermeneutics is an important part of the development of intellectual thought, and it provides important theoretical insights for interpretation in phenomenology; this is because phenomenology delves deep into a real phenomenon, and the researcher's task is to find the meaning of certain manifestations of the phenomenon.

Another important basic theoretical insight of phenomenology is ideography. Ideography is interested in the details of the individual case, which leads to a deep analysis of a specific experience, going into all the subtle manifestations

(Smith et al. 2009). Furthermore, the analysis must be carried out carefully and systematically. The second important thing associated with ideography is that phenomenology tries to figure out how a particular phenomenon is experienced and perceived by specific people in specific contexts; therefore, in phenomenology cases of the analysis of the phenomenon are purposefully selected, and sometimes a separate case analysis can be used. The research is designed so that, having analyzed specific cases, certain generalizations can be made.

Heidegger introduced interpretation as a critical tool to the process of understanding. Claiming that to be human is to be bound to interpret, Heidegger (1972) stressed that every encounter involves an interpretation influenced by an individual's background. In fact, a researcher applying an interpretive process seeks to bring understanding and disclosure of phenomena through language. What is more, hermeneutics is the study of human activity as texts with a view toward interpretation to find intended or expressed meanings (Kvale 1996). Supporting Heidegger's view that language and understanding are inseparable structural aspects of human "being-in-the-world," Gadamer (1999) stated, "Language is the universal medium in which understanding occurs. Understanding occurs in interpreting." Gadamer viewed questioning as an essential aspect of the interpretive process as it helps make new horizons and understandings possible:

> Understanding is always more than merely re-creating someone else's meaning. Questioning opens up possibilities of meaning, and thus what is meaningful passes into one's own thinking on the subject… To reach an understanding in a dialogue is not merely a matter of putting oneself forward and successfully asserting one's own point of view, but being transformed into a communion in which we do not remain what we were. (Gadamer 1999, p. 375)

Gadamer (1999) considered understanding and interpretation as intertwined together, and he also viewed interpretation as an always-evolving process, which makes a definitive interpretation not likely ever possible. He considered that methods of phenomenology are not totally objective and separate from the user, and he expressed his opinion that not only bracketing was impossible but the attempts to manifest bracketing as a method were, in a way, absurd. In fact, he introduced the notion that knowledge depends on the context and background and is continuously evolving in the presence of "historicality" of understanding.

In terms of the phenomenological approach to the research object, it should be noted that the researcher focuses on the respondent's broad experience and its expression in the interviews that best match the content of the experience. The researcher performs the functions of the co-author of the respondent's experience in a variety of ways, stimulating the memories of the past activities, encouraging speaking on the current issues, and so on.

1.1.2.2   Phenomenology on Technology and Education

Gadamer (1999), while analyzing numerous interpretations of the Promethean myth, points out that the Promethean act of giving fire to humans symbolizes human culture based on fire. Fire represents knowledge, the ability to create and the ability to invent new things. The humans in the myth lived in caves and were helpless in the material world before Prometheus brought fire and gave many human arts like astronomy, shipping, medicine, and all kinds of knowledge and possibilities of technological advancement; however, technologies in their own right challenge human existence and in a way disrupt or fragment human life. In fact, the Promethean myth embodies the ability of humanity to create, to invent new things, bringing more comfort into life while at the same time being plagued by their creation like Prometheus is plagued by the mythical eagle constantly eating his liver.

Heidegger (1971), reflecting on things, provides two modalities of things or two essential ways in which we engage with things. The two modalities are using things and thinking about them. In everyday life, we engage in the first modality—using things and perceiving them in a way that takes them for granted—but sometimes we reflect on the presence of things. Heidegger (1971) gives his example of how the nature of a hammer is noticed when the hammer breaks. He speaks about positionality of things, the notion that reveals how the meaning of things is showing and hiding at the same time. Heidegger says, "In all this disguising of positionality, the glimmer of world still lights up, the truth of being flashes." Analyzing the question of technology, Heidegger (1971) reveals that technology is not only tools; he introduces the notion of "ontotheology" of our existence and states that technology enables humans to implement their consuming desires. He warns against the danger of technology, which has the power to shape not only our physical but also our spiritual and social life.

Ihde (1993) states that technologies should be understood not as mere objects and focuses on human–technology relations. The author distinguishes embodiment, hermeneutic, alterity, and background relations. The embodiment relation represents such technologies that became part of our body like body extensions, for example, glasses and clothes. The hermeneutic relation represents things that are used to measure and interpret our world, like microscopes and similar equipment. Alterity relations represent relations with pieces of technology, wherein humans attach personalities to technological devices like cars and computers, where there is a tendency to anthropomorphize the technological objects. The background relations show that some pieces of technology are perceived simply as parts of our environment, like electricity and furniture.

Dreyfus (2012) writes about technologies, criticizing the optimism of the creators of artificial intelligence. Artificial intelligence optimists envision a future when our brains will be hooked to digital technology that will change our cognitive selves; however, Dreyfus (2012) considers such a reaction as the natural initial optimism. He is also concerned about rationalization of human practices like friendship, such as having friends for the reasons of health: "Marginal practices always risk being taken over by technological rational understanding and made efficient and productive... As soon as you have friends for your health or your career you've got friendship, which is of technological–rational kind" (Dreyfus 1991, p. 7). His insights are connected to the meanings of friendship and intimate relations in the environments of social media.

The representative of techno-genetic phenomenology Stiegler (1998) returns to the Promethean myth, trying to show that humanity has been intertwined with technology from the very beginning. The author reveals that Prometheus, by giving fire to humans, in fact gave them the beginning of technology. According to Stiegler's (1998) interpretation of the Promethean myth, because of the fault of Epimetheus humans lack the original property of their existence. Epimetheus, the brother of Prometheus, asked him to let him perform the task of blowing life into mortals, which was given by Zeus to Prometheus. During this process, he gave certain qualities to the creatures and, when it was the turn of humans, there were no qualities left. Then, Epimetheus went to Prometheus to ask him to solve the problem and, seeing that humans—weak creatures—would not survive, Prometheus stole fire from gods and gave it to humans. According to Stiegler (1998), therefore, humans became dependent on artificial means. They invent tools and technology as artificial body parts or prostheses and are doomed to a life of prosthetic beings—cyborgs, intertwined with technology, creating technology and being affected or created by technology. Stiegler (2010) also analyzes the development of technologies of cognition, the development of new media, and the pedagogical meaning of attention, warning against capturing people's attention simply for the purposes of commerce of technical industry.

Van Manen (2014) distinguishes five kinds of human–technology relations, stressing their cyborg nature introduced by Ihde (1993):

- First, the author speaks about experiencing technology as taken for granted as our life is filled with appearing technologies that are designed to make our life easier, more comfortable, and productive.
- Experiencing technology ontically, which means that scholars keep trying to identify how various kinds of digital technology, like computers and the Internet, can be understood.
- Experiencing technology ontotheologically. This is closely related to Heidegger's (1971) insights on technology as always modifying and transforming the world and shaping human experience. Also, based on Heideggerian philosophical insights, Thompson (2005) discusses the dangers and advantages of technology, including educational environments.
- Experiencing technology as technics, which is related to Stiegler's (2010) insights on human relations, with technology revealing the cyborgian human existence.
- Experiencing technology aesthetically. This is related to scholarly studies of sexuality and experience of the aesthetic. Perniola (2004) explores how technology influences human relations and the experience of Eros, sexuality, and the aesthetic of things. Adams (2012) studies how technology influences the student–teacher relationship while adapting new media and technologies in the educational processes.

Arnold (2003) gives a substantial analysis of major theoretical approaches to the functioning of socio-technological systems. The first approach is techno-determinist or substantive (Feenberg 1999), relying on linear causal effects and strong beliefs that technology tools enable specific social conditions. Similar ideas are expressed in innovation diffusion models by Rogers (1995). Next is the social construction approach (MacKenzie and Wajcman 1999), which starts from social conditions and finishes with technology, allowing the existence of a multitude of meanings of technology for different people in different social conditions; for example, technology could be cheap or expensive, controlled by engineers or business people. Both approaches, however, are characteristic of theoretical separation of social condition and technology, which are viewed separately, and then the cause-and-effect rule is applied. The network approach goes further, stating that both technology and social condition comprise at the same time cause and effect (Orlikowski 2000). It provides a further theoretical move, not dividing humans and technology but viewing them in binary connection.

Similar to the fundamental observation by Heidegger that technology does not change the world but enframes the world in a certain manner, humans apprehend the world through a technological frame, and by seemingly answering a need or a question, technology changes the question and the answer at the same time; for example, social networks work not only as means of enhanced communication—they change the understanding of community and communication itself inside the community. Technology acts not only as a tool but on higher metaphysical level. Heidegger observes that by abolishing distance technology did not create the world where humans are closer to each other; actually, distance abolition just led to destruction of closeness. Borgmann (1987) supports Heideggerian ideas by asserting that technology is designed to take the multitude of functions of a process in order to facilitate it; however, finally, technology takes up all the functions and in fact eliminates the process. Ihde (1990) looks at technology through an amplification and reduction framework; for example, when a microscope discloses the world of tiny particles it automatically closes the world of the surroundings in the room.

According to van Manen (2014), the practical orientation of phenomenology and its application to research professional activities in education find their beginnings in the writings of the scholars of the Dutch or Utrecht school such as F. Buytendijk, M. Langeveld, F. Bollnow, and others. Phenomenology is sensitive to personal and social activities in education and embraces the understanding that life and how we experience it are not always logical or rational; in many cases, it demands the uncovering of subtle, enigmatic existential meaning through language and transcendental insight. Phenomenological research addresses a number of educational issues such as learning through the body, writing online, being an authentic teacher, ambiguities in becoming professionals, and schools as places that disturb the self.

Especially, prolific researchers who are working in the application of phenomenological inquiry in educational issues are Max van Manen and Catherine Adams with their recent research on understanding learning and teaching online. Van Manen (2010) explores the pedagogy of social media technologies and how they alter young students' experiences of privacy, secrecy, and solitude. The researcher identifies the effects of pedagogy of social media as Momus effects, considering the ways in which students stay in touch with each other and how they experience their intimacies and inner lives. Adams (2010) researches social media use in university studies and observes that taking ICT into the classroom reshapes teaching relations with students and ways of interpreting the world. Technology in use immediately reforms, deforms, and invites one to conform to the new horizons of our living world. The author frames her paper in light of Martin Heidegger's "Being and Time" (1962) and "The question concerning technology and other essays" (1977), the writings which reveal how new things open new boundaries of our world, give new structures of meaning, and provoke us to a different style of living and thinking.

1.2   The Concept of Social Media

Social media is a rapidly developing application of Web 2.0 technologies, so it is important to overview and discuss the definition of "social media" to identify what is considered to be social media and what are its main functions and features. Additionally, ways in which educational institutions use social media and how social media could be integrated into university studies are discussed further.

1.2.1   Definition of "Social Media"

A "social media" definition requires special consideration as it has not been finalized yet. Social media is still developing, new applications appear, and social media embraces various applications, which themselves undergo development and changes. The numerous applications of social media include such developments as wikis, blogging, social networking, and podcasting. Kaplan and Haenlein (2010) identify six different types of social media: collaborative projects, blogs and micro-blogs, content communities, social networking sites, virtual game worlds, and virtual communities. Various technologies include blogs, picture-sharing, vlogs, wall postings, email, instant messaging, music-sharing, and crowd sourcing. The development of social media causes the ongoing evolvement of the "social media" definition; thus, it is possible to find numerous definitions of "social media" online. According to the *Merriam-Webster Online Dictionary,* social media includes various "forms of electronic communication (such as Web sites for social networking and micro-blogging) through which users create online communities to share information, ideas, personal messages, and other contents (such as videos)."

Boyd and Ellison (2007) define "social network sites" as "web-based services that allow individuals to construct a public or semi-public profile within a bounded system, articulate a list of other users with whom they share a connection, and view and traverse their list of connections and those made by others within the system" (Boyd and Ellison 2007, p. 211).

The above definitions recognize social networking as an essential example of social media; however, social media embraces far more applications than just social networking, and, as has been mentioned above, with the evolvement of social media the applications are developing and changing. Kaplan and Haenlein (2010) identify social media as "a group of Internet-based applications that build on the ideological and technological foundations of Web 2.0, which allows the creation and exchange of user-generated content" (Kaplan and Haenlein 2010, p. 61). The common notion connecting the definitions of "social media" is the mixture of information technologies and social interaction leading to co-creation of knowledge and content. User-generated content becomes the main attribute of social media; in addition, the content could be changed and modified, redefined, and improved by multiple users. Technology permits its users to interconnect in the process of content creation and, by using multiple channels, steadily modify and change it. Social media extends the range of traditional media, which is based on television, radio, newspapers, and other printed publications. Both traditional and social media provide the possibility of reaching wide global audiences; however, social media does it almost without any cost or special resources. Social media has an inherent democratic nature, which allows its users to actively participate without any special training or qualification or even permission to publish their material. Immediacy and recency are supplementary traits differentiating social media from traditional media. Users of social media can publicize their events immediately and comment on them, they receive instant reactions, and the process of creation is in constant continuation. Refashioning is another attribute of social media, while traditional media is complicated to change once it has been published. Social media enables the participants to comment on, change, and rearrange its content. Social media provides a publicly open space where people can exchange their ideas and opinions.

Kietzmann and Hermkens (2011) identify certain functional blocks in the framework of social media: "identity," which represents how users create their personal representations; "conversations," which show how users communicate and converse with others; "sharing," which reveals information bits that users introduce and share with others at the same time, allowing other users to modify the shared content; "presence," which represents the accessibility of users to others; "relationships," which show how users relate to each other; "reputation," which primarily represents trustworthiness of users; and "groups," which reveal user-constructed communities or subcommunities. The defined framework includes functional blocks such as conversations, sharing, and relationships, which represent important aspects in study environments. Learning paradigms in educational contexts require collaboration and sharing, which enable students to actively construct and apply their own understanding and knowledge. In fact, user-generated content and its collective active use and redefinition are the main feature characteristics of all social media applications, which is of the utmost importance in study environments as it ensures active learning and knowledge creation.

1.2.2   Social Media in Education

As McLuhan (2003) states, people are creating technologies and then technologies are creating people. People become what they are looking at. While speaking about media, McLuhan has in mind means of information and communication—means that have multiple channels, carry multiple meanings, and create multiple intersections of meanings and messages, thereby constantly generating new messages, new meanings, and new knowledge. Due to the development of technologies, media evolved into social media, which itself is in the process of ongoing change and constant development. It allows multiple interactions among media users without any limitations of space in real time. The emergence of social media and its rapid evolvement is closely connected with the development of Web 2.0 technologies. LeNoue et al. (2011) describe Web 2.0 in the following way: "These applications have provided Internet users with the ability to easily create, contribute, communicate, and collaborate in the online environment without need for specialized programming knowledge" (LeNoue et al. 2011, p. 5).

Technological change has influenced education in general in recent years. As researchers identify (Glastra et al. 2004), information technologies—including social media technologies—have blurred the boundaries between home, work, leisure, learning, and play, and they have reshaped our lifestyles and social interaction. University studies have also been influenced by social media. According to Castells (2007), a knowledge economy is our best route for success and the Internet is dissolving physical barriers; its technologies like social media are changing the ways in which people live, learn, and study. Social media is pervasive in all spheres of our life, and it naturally enters the education arena, which poses the question of an effective social media use in education. Rapid development of Web 2.0 technologies and their applications such as social media reveal that it is important to construct understanding about the effective use of these technologies in educational processes (De Rossi 2007). The numbers of social media users are rapidly increasing, and the future generation of students already actively participates in social media and started creating digital content.

Various social media technologies and their applications, enhanced by Web 2.0 possibilities, are gaining more and more importance because they stimulate digital literacy and act as effective means of teaching and learning in education. Educational institutions have the possibility of applying Web 2.0-provided advantages like simple and fast creation of micro-content and social factor of communication, which means instant communication and feedback. Both simple and fast content creation and social factor of communication, in turn, promote further creation and development of digital content, at the same time providing opportunities for improving communicative skills, which are very important in the study process (Scardamalia 2002). Educational institutions have the opportunity to use social media in the study process, and a number of institutions started using it. Obviously, different institutions choose different ways of using social media. Some institutions attempt to create a safe study environment, so they choose to build their own internal social media networks; other institutions choose an integrative outlook, use already existing publicly available social media, and try placing the study process into the public space by creating studying communities there. Bearing in mind the versatility of approaches, it is important to gather and compare diverse insights and the experiences of numerous authors on using social media.

Theoretical research on social media use in education, practical experience, and insights of various authors permits us to foresee and introduce successful and effective scenarios of the technology use. This is especially important in bridging formal and informal education (Burkšaitienė and Teresevičienė 2004). Social media could be used as an element of informal education in the process of formal studies, developing student creative collaboration in an informal medium. If we reflect on how collaboration happens in the real-life situations, we will identify that, first, students consult each other, then they discuss, and finally they creatively apply the new ideas by creating new content and new knowledge; what is more, at the same time they learn from each other during the whole process. The goal of learning is knowledge preservation, transmission, and creative application, so we anticipate that the students will acquire the new information during the studies and apply it accordingly later.

During the last decades, assessment and recognition of non-formal and informal learning at university have become a reality in many countries worldwide, including European Union countries where methodologies of assessment of learning outcomes gained through different non-academic learning environments have been created. Concurrently, systems of identification, assessment, and accreditation have been introduced at all levels of education, and methods of assessment have been established. In Lithuania, assessment and recognition of non-formal and informal learning at university have made its first steps. Unlike students who traditionally enter after finishing secondary school, adult learners applying to a university possess learning outcomes gained through different learning environments. Thus, a new role for universities emerges, i.e., to make their shift from one sector of education to another smooth, to bridge the world of work and the university by providing access to higher education, and to establish a procedure of assessing and recognizing this kind of learning. As it is reported (Burkšaitienė and Šliogerienė 2013), this proves to be a real challenge due to several reasons. First, universities have to make an important decision before adopting a procedure of assessment and recognition of non-formal and informal learning, i.e., to acknowledge that learning is a value that does not depend on the environment in which it occurs and that learning gained through non-traditional study routes may be equivalent to learning acquired at university. Second, members of academia have to discuss a particularly important issue—how does one assess learning that occurred outside the institution without diminishing the quality of university studies, or in other words, how does one ensure that learning which is recognized is in fact equivalent to academic level learning? Thirdly, the decision requires thorough preparation and resources, including human, financial, infrastructural, and time resources necessary for the creation and maintenance of an administrative unit that is responsible for the implementation of the procedure, creation of support systems available for adults at university, for establishing what is going to be assessed, setting assessment criteria, training consultants and assessors as well as informing the public about the new possibility that is available to adult learners. Fourthly, assessment of non-formal and informal learning at university is a new and specific task, because traditionally learning assessed at university has been provided to students by teachers at university, whereas learning that occurred

outside the university has never been organized according to university study programs, and its outcomes are not always easily defined (Boud 2001, cited by Costley
and Armsby 2007). Finally, it is likely that adults who have never been members
of the student community will not have acquired the academic language typical to
the study subject, which may lead to a difficulty in translating learning outcomes
into academic discourse (Peters 2005). Taking all these challenges into account,
we see that new forms, methods, and education platforms have to be organized. In
Lithuanian higher education institutions, the learners are given a considerable degree
of autonomy (attendance is not mandatory, there is a great emphasis on individual
work on their curricula, study materials are available online, and such); besides,
many students have part-time jobs. Thus, being a self-regulated student should be
of great importance to manage the provided autonomy and the degree of responsibility one has to take for one's studies. Self-regulated learning (SRL) has been
widely discussed and researched from the point of view of students' perspective
to develop self-monitoring skills, increase motivation, register their own progress,
from the point of view of a teacher change the role in the assessment moving from
summative to formative, and provide frequent continuous feedback in order to help
learners to self-regulate their academic achievement process. SRL is also considered
to be as a key to success in a career (Boekaerts 1999). Having gained self-directing
skills at the university employees can deal with problematic issues on their own and
become the type of professionals many organizations seek because of their ability
to adapt to a changing environment. Has a teacher's role changed due to a different organization of studies and the amount of content a student can access on his
own? A teacher in front of a group of students is no longer treated as the principal
source of information. Following a student-centered approach, a teacher's role has
shifted: He/she is a facilitator, a mentor, and in some cases a supervisor. The shift
was inspired by the usage of new teaching/learning methods, which helped to better
develop particular skills. It is apparent that contemporary teachers face the reality
that most of their instructions tend to focus on content knowledge and not on the
process of learning the transferable reflective skills. Reflective students are likely to
be more self-aware and self-critical; honest about themselves, and open to criticism
and feedback; objective in weighing up evidence; open to, and prepared to try, different approaches; and curious to discover other approaches, motivated to improve,
and more able to carry through independent learning. It is a challenge to provide the
correct amount of guidance without giving too much direction. Direction is needed to
help students identify areas of difficulty, but too much guidance detracts from their
sense of ownership of the learning project. A teacher has to reduce direction and
support gradually as the student gets more mature and self-confident. Teachers need
to develop a class structure and teaching style that encourages creativity, reflective
thinking, and self-directed learning. It is important that teachers enable students to
have the freedom to ask questions and take intellectual risks in their written assign-
ments and discussion groups. A teacher needs to restructure the teaching/learning
process to autonomous studies where a student gradually becomes a self-directed
learner. Learners start using different sources of information, and formal learning is
no longer the only way of developing one's competences.

Social interaction promotes the process of learning (Paiva Franco 2008). Collective work on learning tasks enables the students to accommodate different attitudes, to negotiate and discuss, and to achieve consolidated understanding. Common aims increase the process of learning, engaging the application of conceptual knowledge in solving essential problems. The activities of formal education could be successfully related to the activities on social media. Social media enables the use of external sources and also gives grounds for relocating teaching/learning from formal settings into informal situations of real communication on social media. Learners have chances to know each other informally, and sometimes student hobbies and supplementary knowledge provide a better approach to tasks and facilitate the study process. Social relations can provide the grounds for informal learning when there is a possibility to get consultations from one's social relations. Using social media also allows the creation of a learning community online as an addition to classroom activities.

New Web 2.0 technologies revealed the means of modern forms of communication and creative collaboration on the Internet, where creative collaboration describes a relationship between people with a common purpose of creating new objects through certain ideas and shared understanding of something new. Social media is considered to be an exceptional example of the application of new technologies. It acts as a driving force itself because it was created outside educational institutions. Many education practitioners and experts assume that the use of social media in the study process poses an essential challenge to educational institutions by raising questions of how to use social media and integrate it into sustainable study process. Education experts admit that social media can have a great influence on education by relating and integrating formal and informal learning (Burkšaitienė and Teresevičienė 2004).

Tools of social media could also be used for developing skills of creative collaboration by using project work methodology and integrating social media tools into project work. Creativity in the learning process means an innovative application of new knowledge for the learner, and it becomes a transversal skill in educational processes (Scardamalia 2002; Chai and Fan 2018). This skill is one of the most important in the study process since it requires the highest degree of thinking and stimulates the learner to find new ways of learning; thus, using social media tools for problem solving and project work, students employ a high degree of creativity, which could be incorporated into the successful study process (Scardamalia 2002; LeNoue et al. 2011). Another feature, namely contacting experts, peers, and additional sources of information, draws more interest and value into the learning process. Also, learners can publicize their work in the public Internet space, and this increases students' understanding of the importance of their work in the wider world. In such a way, learners can compare their work with peer achievements in a broader context. They have the possibility to experience what their work looks like in the context of the study subject, and this gives additional value to the study process. While using social media, real information exchange is promoted as well as a higher level of collaboration and creativity (Scardamalia 2002; LeNoue et al. 2011).

Recently, there have appeared many discussions on the use of social media in education (De Rossi 2007). Although the use of social media in the study process comprises many positive aspects, on the theoretical level there still exist some essential questions. Traditional education organizations are highly structured concerning the content and study process, while social media use demands less control of the user-generated content. Such a situation poses a challenge to the traditional perception of the effective teaching/learning control: "As it could have been expected, traditional academic institutions usually have a suspicious view towards social media existence in the lives of their students, however, now they have to review all the new aspects and consequences of new technology generated communication which is so popular among the young generation" (De Rossi 2007, p. 4). Social media use actually poses the questions of how the process of managing freely generated digital content could be integrated into the traditional teaching/learning environment, and whether the skills gained through social media use blend into the traditional teaching/learning environment.

In fact, social learning theory is not a novelty; its foundation approach recognizes that social communication is the grounds of the effective learning because learning does not happen in isolation. Paiva Franco (2008) states, in his article about e-learning, "Social interaction promoting cognition is essentially important to ensure the learning process either in real or online auditorium." This approach is supported by the Vygotskian proximate zone development theory, which Paiva Franco (2008) relies on, stating that the "distance between what the students are able to achieve themselves and what they can achieve with the help of others allows us to understand how socio-constructivist environment can provide productive learning possibilities" (Paiva Franco 2008, p. 3). Students communicating and learning in a digital environment actually profit from communication with both their peers and lecturers who enable the students to attain a higher level of understanding. Online discussions enable students to enrich their study experience; however, the question arises as to whether social media can secure the appropriate context in which skills of social communication could be developed. The present research also focuses on two theories—the key functions of verbal communication and the perception theory—in order to reveal how social media is perceived and communicated to the audience of different cultures. Communication, according to Hunt et al. (1998), serves a few functions, some of which are the key ones of verbal communication, namely:

- Communication helps people define reality.
- Communication helps people organize complex ideas and experiences into meaningful categories.
- Communication helps people think.
- Communication helps people shape their attitudes about their world.

Taking these functions into account, the questions of how people perceive the defined reality, how they are capable of organizing complex ideas, how attitudes are being influenced by their understanding, and the complexities of the language signs and symbols should be carefully considered. Verbal communication is a message that could be sent via language; it might appear in different forms—a face-to-face interaction, by phone, via email, Skyping—using any social media as long as language is involved. The way we decode a message would design our thinking, interaction, and perception of reality.

According to the perception theory, perception is a process through which knowledge of the objective world is acquired; therefore, it can reveal how this interaction is perceived (Maund 2003; Freitas 2014 cited in Burkšaitienė 2017). The degree of perception of verbal expressions in communication depends on a great variety of factors: physiological, psychological, and intellectual abilities, language, and cultural awareness. The degree of perception fosters creative interpretation and inspires a decoder to look at the meaning in terms of creativeness and higher awareness of a particular culture.

De Rossi (2007) discusses the new applications of social media in education and recognizes that students use social media for communication, collaboration, and creation of digital knowledge. As has been discussed above, learning is studying from not only textbooks but also communication and interaction. Students are able to use social media for completing creative projects and achieving good results with them. Before the application of digital technologies, and especially social media, communication in learning had been limited to the physical space of a classroom; however, social media has enhanced the space available for the social component in learning. The most usable social media platforms in education include blogging, Facebook, Twitter, LinkedIn, and Google+. The use of social media in education demands taking into account the interests and needs of the students and also broadening connections with the students beyond physical presence in the classroom.

According to Marquis (2012), social media is used by educational institutions in certain ways (see Fig. 1.1):

- Classroom resources include: (a) sharing announcements and discussions, and (b) various sites on which to access important resources and search for information.
- Institution image comprises: (a) event popularization, (b) attracting participants, and (c) developing the institution's representation.
- Reaching out for future students includes: (a) information on the institution, (b) virtual tours, and (c) various videos.
- Professional development consists of: (a) information on the activities inside the institution and (b) collaboration with professionals from other institutions and networking with other professionals.
- Staying connected means: (a) further interaction through various groups, blogging, conversing, exchanging ideas, and getting feedback, (b) interaction with the community, various businesses, and other organizations, and (c) connecting alumni groups.

Summarizing the schematic representation of social media use by educational institutions, two main trends could be distinguished: firstly, social media use for learning—both by students and teachers—for their professional development and, secondly, representative–communicative use of social media to promote the institution and keep wired into the community.

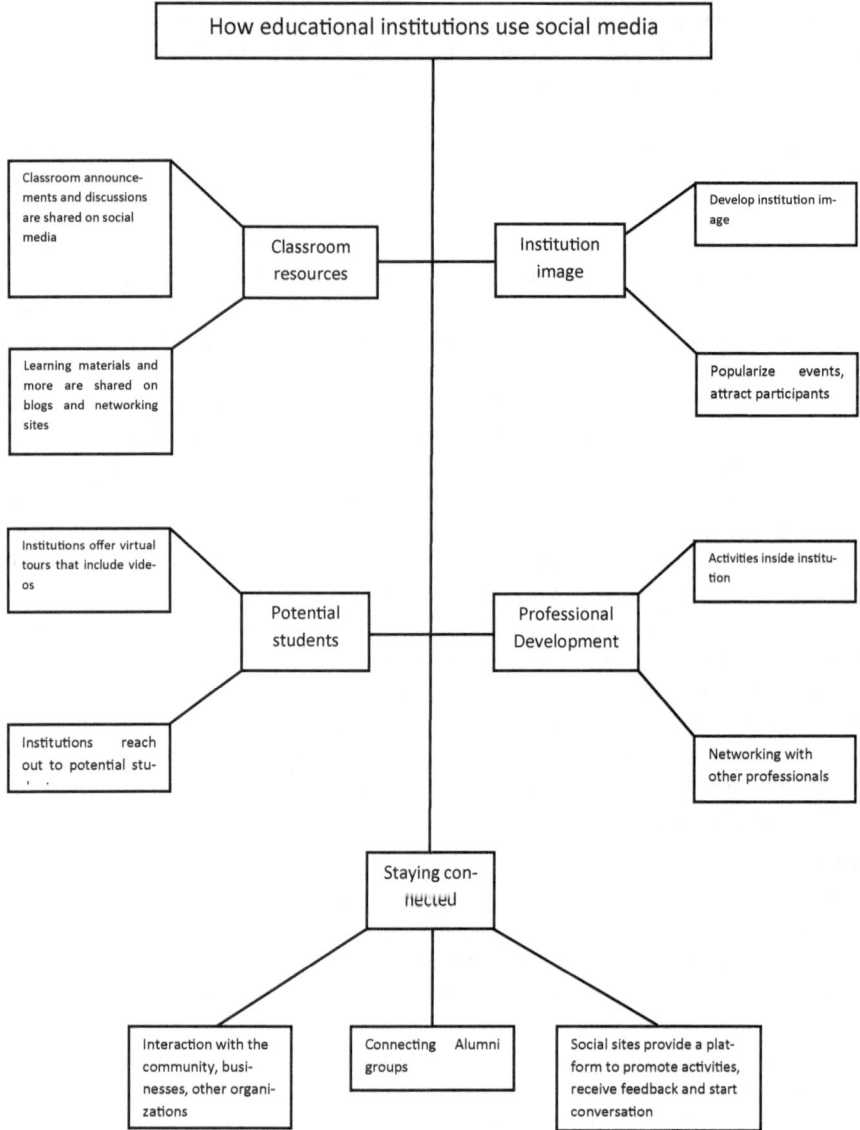

Fig. 1.1 Social media use in educational institutions (modified from: Marquis 2012, pp. 1–2)

1.2.3 Integrating Social Media into the Study Process

Social media use in the study process demands careful contemplation from the teaching professionals. It is necessary to establish the relevant tools to be used for teaching and learning purposes, which requires the overview of what the study process consists of and what is being done, how social media could improve and expand teaching and learning, and how it could be applied. Along with the use of appropriate tools, creating a learning network to keep the study participants informed as well is important. Bernoff and Li (2008) identify the key points teaching professionals should take into consideration:

- The needs of the audience, and what additional assistance with social media technology a need might exist for;
- The aim of teaching and learning—what is required to be achieved is considered and then social media tools are chosen;
- The objectives—how information is going to be presented, how teacher/student and student/student communication is going to be enhanced, and how student content creation and use of social media are going to be supported;
- Strategies—how social media is going to be used in teaching and learning (e.g., collaborative writing, multimodal communication, or online networking);
- A tool or technology of social media (e.g., a wiki, a community, or blogs)— usually after the audience has been identified, the aim, objectives, strategies, and technology are chosen.

Teaching and learning objectives are of key importance for educators. Bloom's taxonomy identifies a classification of teaching and learning objectives widely used by educators. Knowledge is connected to the cognitive domain of the taxonomy, which is segmented into six levels. It is also stated that learning at higher levels depends on the previous knowledge acquired at lower levels.

The first level is related to information and facts, generally called "knowledge." The next level introduces comprehension—the ability to understand facts and to compare or interpret them. The third level embraces the use of the gained knowledge, for example, solving problems through knowledge application. The fourth level is the analysis level, which includes the ability to break information into parts, examining causes and effects. After that, the fifth—synthesis—level is introduced, which includes compiling information together and combining various elements of information into new patterns by creating alternative solutions. Finally, the sixth level is the evaluation level, connected to making judgments about information and validity of the ideas. Bloom's taxonomy is presented in Table 1.1.

Bloom's taxonomy was reconsidered in 1990 and published in 2001 (see Table 1.2). It introduces the use of verbs rather than nouns in the categorization and features a rearranged sequence of the categories, which is organized into increasing order, renewing the category of highest position. In the revised Bloom's taxonomy, creativity is considered to be higher in the cognitive domain than evaluation.

One of the essential reconsiderations of Bloom's taxonomy was the naming of the levels by verbs representing actions:

- Creating—designing, constructing, planning, producing, inventing, devising, making;
- Evaluating—checking, hypothesizing, critiquing, experimenting, judging, testing, detecting, monitoring;

Table 1.1 Bloom's taxonomy

	Levels	Verbs
Highest	• Evaluation	Assess, grade, recommend, decide, test, convince, support, measure, select, conclude
↑	• Synthesis	Combine, rearrange, create, rewrite, design, integrate, compose, modify, plan, invent, generalize
↑	• Analysis	Analyze, explain, arrange, select, separate, connect, infer, order, classify, compare, debate
↑	• Application	Apply, demonstrate, complete, illustrate, show, solve, examine, modify, relate, change
↑	• Comprehension	Explain, discuss, compare, interpret, describe, contrast, outline, restate, summarize, distinguish
Lowest	• Knowledge	List, define, tell, describe, identify, show, collect, quote, name

Table 1.2 Bloom's revised taxonomy

Higher-order thinking skills	Creating
↑	Evaluating
↑	Analyzing
↑	Applying
↑	Understanding
Lower-order thinking skills	Remembering

- Analyzing—comparing, organizing, deconstructing, attributing, outlining, finding, structuring, integrating;
- Applying—implementing, carrying out, using, executing;
- Understanding—interpreting, summarizing, inferring, paraphrasing, classifying, comparing, explaining, exemplifying;
- Remembering—recognizing, listing, describing, identifying, retrieving, naming, locating, finding.

The verbs represent the action and the activities performed in the educational processes, and they provide a clear representation of learning objectives; however, the use of social media in education requires reconsideration of the objectives and activities already relying on the framework of social media. Bloom's taxonomy was therefore digitized, and a new model of digital taxonomy was suggested by addressing the new aspects of social media integration into education (adapted from Skiba 2013, 2).

We modified the pyramid into the following table (see Table 1.3), so that the newly created digital taxonomy could be more easily compared to Bloom's taxonomy by incorporating some of social media tools that suit the taxonomical level, and so that it could be used in educational activities. The table could provide a better idea of how social media tools could be chosen while pursuing certain learning objectives.

Adaptation of various social media tools to the teaching and learning objectives of Bloom's taxonomy establishes considerable grounds for social media integration

Table 1.3 Social media tools based on Bloom's revised digital taxonomy

Levels	Tools
Creating	Prezi for preparing presentations Wiki spaces for collaborative contributions *Keywords: adopt, compile, compose, create, design, generate, invent, make, model, organize, plan, portray, publish, revise, rewrite, synthesize, write*
Evaluating	SurveyMonkey for making surveys, evaluating opinions YouTube for evaluating ideas *Keywords: assess, compare, consider, contrast, critique, debase, evaluate, explain, interpret, justify, prioritize, prove, recommend, relate, support, test*
Analyzing	Exploratree for building portfolios of useful thinking guides, analyzing the different perspectives *Keywords: analyze, classify, compare, contrast, correlate, differentiate, examine, group, identify, order, outline, select, sequence, sort, survey*
Applying	Wikipedia for choosing and adapting information *Keywords: adapt, choose, construct, determine, develop, organize, practice, predict, present, produce, select, show, sketch, respond*
Understanding	Bubble us for creating mind maps The Periodic Table of Videos or watching and understanding *Keywords: compare, conclude, contrast, define, describe, estimate, explain, identify, interpret, paraphrase, summarize, understand*
Remembering	Delicious for social bookmarking Flickr for sharing photographs *Keywords: define, describe, find, identify, label, list, match, name, select, show, tell, write*

into the study processes. There are multiple reasons for integration of social media in education. First, social media is free of charge and is widely used by multiple audiences. Next, social media tools fit into the whole range of teaching and learning objectives of Bloom's digitized taxonomy and could be used successfully in developing students' higher-level learning skills, such as creativity and collaboration. In university studies, which are an important part of the broad education scale, social media meets various needs of students in their personal and study lives, and it improves digital and social media skills necessary for employment in the modern information age. Finally, social media steadily enters the university study domain and modifies the character of education itself (Bernoff and Li 2008).

Jucevičienė et al. (2010) distinguish a twofold role for social media in university studies. Firstly, social media itself is perceived as a learning object, which means that teachers and students learn about social media and the learning possibilities provided by social media, and they acquire the necessary skills of how to use social media. Secondly, mastering social media technology provided the opportunity for teachers and students to use social media as a learning tool. It means that social media is applied to enhance teaching and learning and could be defined as a teaching and learning tool and method. In such a way, social media becomes an education technology. The main requirement for university studies is to satisfy student needs and study potential. Social media use in university studies allows wide use of multiple channels of acquiring information using visual, textual, and sound information; it also allows transition from traditional lecturing to individualized active and effective teaching and learning. According to Selvyn (2012), social media as a teaching and learning tools and method allows students to manipulate and organize the study content and learning as well as allows students to establish their own approach to the studied subjects.

There is also a broad approach of technology-enhanced learning (TEL), which analyzes technology (including social media) application in university studies. According to Kirkwood and Price (2014), the term "technology-enhanced learning" is used to describe the use of information and communication technologies (ICT) for teaching and learning. The approach deals with a broad technology application and looks for pedagogical solutions for curriculum design based on TEL. Kirkwood and Price (2014) suggest the potential benefits that TEL might bring into the university study arena. The benefits include: efficiency, which means that educational processes are carried out in a more effective way; enhancement, which means the improvement of the existing educational processes and their outcomes; and transformation, which means radical, positive change in the existing educational processes or even introduction of totally new processes. Transformation appears to be the key expectation, but, as the authors admit, it is not always the case; however, it should be admitted that TEL is a promisingly developing research field.

To conclude, there are emerging ways of integrating social media in university studies. A variety of social media tools could be used for achieving certain learning objectives.

1.3   Social Media and Emerging Changes in University Studies

University studies undergo certain changes related to both technology and society development. In this section, the background information on modernization of university studies is overviewed. It is stressed that university studies are getting closely related to technology application in education where technology acts as a driving force. Also, inner processes in university studies induce the application of the newest technologies to satisfy the demands of appearing needs and changes in the dynamic of university studies. Various forms of social media and course management systems (CMSs), which have certain features of social media, are discussed, disclosing media choices according to learning objectives.

1.3.1   Modernization of University Studies

Information technology, fueled by its rapid development, is becoming pervasive in all spheres of life and is acting as a driving force in changing our life realities. Information technology becomes an active agent, having the capacity within itself to enhance the change. To quote McLuhan (2003) regarding his insights on media: "A medium of communication is not merely a passive conduit for the transmission of information, but rather an active force in creating new social patterns and new perceptual realities" (McLuhan 2003, p. 2).

Drori and Meyer (2006) discuss scientization of modern life, providing an example of how science enters domains such as even religion. The authors present a discussion on the problem of the invention of the modern Sabbath elevator, which works continuously during the Sabbath by opening and shutting its doors at every floor and thus does not require activation in the form of pushing the button. The religious authorities relied on the scientific conclusions, and in fact, the reliance on science has become a routine. There is the semi-academic Institute for Science and Halacha, where rabbis regularly consult with the scientists. And the process of scientization even moves forward in the decision-making sphere; the same example could be quoted, wherein the scientific response was later embedded in the state law and its enforcement mechanisms. In July 2001, the Israeli parliament passed the "Sabbath elevator law," which requires the new elevator as a standard in all high-rise buildings, regulates the cost-sharing procedure among the building occupants, and offers a legal enforcement mechanism for the possible violations of the law (Drori and Meyer 2006). In fact, what the society is facing is the close blend of intertwined science, technologies, and education in order to promote an even more rapid technological advancement and focus on education toward creation of knowledge, newest technologies, and science promotion.

It should be noted that in some international contexts the terms "university studies" and "higher education" are used interchangeably as higher education is ensured by university studies. In Lithuania, however, we have a binary system that allows equal functioning of university and non-university studies. Despite international differences, all international processes taking place in the international higher education arena equally affect university studies. The implementation of technology-driven processes is becoming crucial to most universities (Daniel 1998) as it promotes the process of higher education modernization. Bach et al. (2007) provide an extensive study of how the growth of the Internet influenced higher education and university studies, identifying three main challenges posed by the development of information technology:

1. Availability of quality information and suitability of information systems;
2. Ensuring knowledge management and creation remains the main focus in the era of information technology dominance;
3. Advancement of IT and information-processing skills.

New technologies require new skills to operate them, and often the staff are responsible for educating themselves and equipping themselves with the newest skills, which really causes stress and adds to the perceived workload (Mason and Rennie 2008). In addition, in some cases the availability of technology appears scarce and the staff cannot see the sense of learning skills that are not going to be applied. In 2003–2004, when a survey was carried out to explore the state of teaching English in Lithuanian educational institutions there were cases when technology was available at an educational institution but just in an administrator office or for IT specialist purposes (Antulienė et al. 2005). One of the authors remembers 2007—her last year of teaching at a high school before moving to university. The headmaster was dissatisfied with the lack of multimedia-managing skills among the teaching staff, while the only one multimedia set was available in the whole school. It meant very limited possibilities for regular multimedia use. When she started working at a university where each room was equipped with multimedia and mastering its use was a natural and short process, she realized that availability of technology enhances the acquisition of necessary skills for its operation which is also identified by Venkatesh et al. (2003) that technology availability is a necessary although not sufficient condition in the acceptance of technologies.

Bach et al. (2007) foresee an extensive growth of technology use and online learning in higher education due to the following factors:

- Rapid technological change, which naturally pervades all spheres of life, including education;
- Availability of online technologies due to the fact that the provision of technology is becoming more and more accessible;
- Changing students' lifestyles so that they acquire a part-time job alongside their studies due to the economic and social conditions;
- Increase in student IT skills as the new generation demonstrates an inclination to acquire the skills more easily;
- Growth of higher education to satisfy a growing demand for mass higher education;
- Growth of higher education in the global market fueled by the demand for higher education;
- Globalization processes and the need for international study experience, cooperation, and global networks.

The Bologna Declaration of the European Union promotes international cooperation between the institutions of higher education. The 1999 European ministers' agreement delineated the framework of standardization of higher education processes across Europe and encouragement of international cooperation between universities. In the Bologna Declaration, ministers affirmed their intention to:

- Adopt a system of easily readable and comparable degrees;
- Implement a system based essentially on two main cycles;
- Establish a system of credits (such as the European Credit Transfer and Accumulation System);
- Support the mobility of students, teachers, researchers, and administrative staff;
- Promote European cooperation in quality assurance;

- Promote the European dimensions in higher education (in terms of curricular development and interinstitutional cooperation) (*Focus on Higher Education in Europe* 2010).

The document outlines the elements of an integrated educational system through-out Europe and stresses the importance of cooperation between educational institutions, especially between the institutions of higher education. An extension—the Leuven document—was later added, envisioning student mobility and student-centered learning.

The Leuven/Louvain-la-Neuve Ministerial meeting, held on April 28 and 29, 2009, acknowledged the achievements of the Bologna Process and laid out the priorities for the European Higher Education Area (EHEA) for the next decade:

- Each country should set measurable targets for widening overall participation and increasing the participation of underrepresented social groups in higher education by the end of the next decade.
- By 2020, at least 20% of those graduating in the EHEA should have had a study or training period abroad.
- Lifelong learning and employability are important missions of higher education.
- Student-centered learning should be the goal of ongoing curriculum reform (*Focus on Higher Education in Europe* 2010).

The importance of employability, lifelong learning, and an approach focused on student-centered learning clearly stand out in the document as stepping stones in the process of higher education shift.

Universities are influenced by common logic of mass higher education, which means that as organizational units universities are becoming inclusive and flexible organizations, which for European universities is commonly expressed in the Bologna Declaration. Ramirez (2006) presents a university as a rationalized organization, which means that like any other organizations universities acquire managerial discourse, which is pushing out old socially related values of human development and fulfillment. According to Ramirez (2006), however, who are optimistic, universities have human development goals; they still operate in the knowledge production, transmission, and preservation sphere. The author believes in the development of a socially embedded university that is characterized by broad inclusiveness, social usefulness, and organizational flexibility, as the main tendency of higher education for all promotes the establishment of a user-friendly university.

Cowen (2013) gives an extensive overview of the development of the British universities as well as touching on the world tendencies. The author relates that in the mid-1970s higher education moved toward mass higher education, and around the 1980s there was another move toward effectiveness and efficiency, and universities were forced into the marketplace. Finally, the author characterizes the present day's globalized and technologically equipped world, where universities undergo the processes of internationalization operating in the global markets and embracing technologies. The problems named by Cowen (2013) concern the commoditization of knowledge, students viewed as "consumers" and learning moving toward training of a skilled labor force—so-called mass amateurization. What is more, the author relays data from the Organisation for Economic Cooperation and Development (OECD), the World Bank, and *The Times* ranking data, establishing five dominant trends fueled by globalization and internationalization of universities. These dominant trends (massification, privatization, governance and accountability, international mobility, and ranking and world-class university) represent major innovations in response to market competition and also pose challenges to university studies.

The university, which was protected by numerous theories claiming it as a special place with academic institutional autonomy, providing "academic freedom" for individuals, has undergone changes (Rothblatt 1997). Higher education is heavily influenced by manifold societal factors and decisions, and still, amidst all the influences and changes, higher education is trying to assert its unique role and vision. So what could they be? Burton (1983) states that, traditionally, university implemented three main goals:

- To accumulate, preserve, and disseminate the accumulated knowledge and wisdom;
- To implement practical knowledge in a modern way and develop skills, providing preparation for professional work life;
- To create the future of society by encouraging unrestricted scientific research and experiments.

The development of modern knowledge, skills, and professional qualifications could be adequately implemented if a university really perceives and understands the societal changes and is able to construct a symbiotic educational and research system. This can be done if knowledge, like Stehr (2005) views it, is not perceived and treated as just an accumulation of encyclopedic or academic information or a productive force or just a commodity—if knowledge is perceived as an active agent having the power of reshaping and creating reality. In 1989, Oakeshott addressed university education, stressing that pursuit of learning is not just seeking for a higher place or status or a certificate; it is an activity broadening one's horizons and equipping oneself with readiness to perceive and create new realities.

> The pursuit of learning is not a race in which competitors jockey for the best place, it is not even an argument or a symposium; it is a conversation…. One may go to some sorts of art schools and be taught ten ways of drawing a cat or a dozen tricks to remember in painting an eye, but the scholar as teacher will teach, not how to draw or paint, but how to see. (Oakeshott 1989, p. 97)

In this way, university education or higher education should be distinguished from simple vocational preparation or acquiring a profession or certain qualifications.

> A university will have ceased to exist when its learning has degenerated into what is now called research, when its teaching has become mere instruction and occupies the whole of undergraduate's time, and when those who came to be taught come, not in search of their intellectual fortune, but desire only a qualification for earning a living or a certificate to let them in on the exploitation of the world. (Oakeshott 1989, p. 98)

The question of teaching and instruction is essential in education. Lagemann (2000), while analyzing the history of American education, states that while J. Dewey was trying to put forward his ideas on a holistic approach to education his opponents were trying to improve instruction; specifically, E. L. Trondike worked on instruction improvement using the latest methods in psychology. The difference between instruction and teaching is essential. Teaching allows the existence of multiple perspectives, not looking for the final solution, while instruction provides the answers to all the questions and the effectiveness of instruction could be measured. In fact, according to Lagemann (2000), the prevalence of instruction means the triumph of a mechanistic approach that uses behaviorist categories to explain life phenomena. It as well leads to what Weber (1946) analyzed in his works on bureaucracy of education, pointing out that a specialized exam system is directed to gradation, certification, and preparation of a specialist rather focusing on shaping an educated person.

The ideas of a deeper outlook on university education have been supported and developed by contemporary authors. "University" cannot simply be defined as knowledge acquisition and preparation for professional activities. According to Barnett (1990), university is a personality building and development process directed toward purposeful human transformation; it is more a self-formation process through critical thinking directed inward and inspiring personal transformation when the learner works independently and takes decisions independently. University should provide its students with a broader outlook and the opportunity to be able to not only identify strengths but also see the limitations and be able to apply critical thinking. As a concept, critical thinking is defined by The U.S. National Council for Excellence in Critical Thinking as an intellectually disciplined process of actively and skillfully conceptualizing and analyzing, or evaluating information obtained by experience, reflection, or communication, which guides human beliefs and actions. In fact, the essence of the approach to university as shaping an educated person and providing intellectual freedom has remained the same; contemporary scholars just try to analyze the dominant trends of mass higher education and new challenges. Zusman (2005) identifies the fact that universities undergoing internationalization, embracing technologies, being pushed into the marketplace, and being ranked by external evaluators face the main challenge of preserving their focus on shaping an educated person. As has been mentioned above, university education is heavily influenced by various societal and technological factors, and the appearance and application of social media induce certain shifts and reflections on the ongoing processes.

1.3.2   Learning Management Systems in University Studies

Concerning the certain overlap of the terminology between learning management systems (LMSs) and CMSs, it should be noted that confusion appeared due to the similarities of the definitions; however, according to Carliner (2005), LMSs were designed for workplace learning environments while CMSs were designed for academic purposes. The author stresses the main difference as the fact that LMSs provide a more personalized approach as they offer catalogues of courses and track learner participation. Petherbridge (2007), however, notices a certain blur between the terms due to the technological move toward a more personalized learning, and thus he identifies the tendency to use the more inclusive "LMS" term. Currently, many universities use LMSs such as Blackboard, Sakai, and Moodle. They are used to present course materials and to ensure interaction between teachers and their students and between peers. LMSs are stated to be compatible with student-centered approaches based on constructivist theories. Students are enabled to discover and work out the concepts and knowledge using critical analysis and reflection (Mason and Rennie 2008). Multiple information resources on the Web, peer discussions, tutor guidance, and collaborative activities combined together allow individual students to construct their knowledge of a subject. M. Dougiamas, the creator of Moodle, suggests five social constructionist features reflected in Moodle (Dougiamas and Taylor 2002):

1. In a really collaborative environment, the participants of educational processes become both teachers and learners. The activities in Moodle (e.g., wikis, forums, messaging, databases, and glossaries) allow the users to actively control the shared content within the framework of the courses. Students have the opportunity to act as instructors for other students by sharing and adding their course experiences. The instructor's ability to arrange different groups of students to be responsible for sharing may blur the differences between student and teacher roles.
2. Learning is stimulated by creating or expressing something for others. Moodle has various tools for sharing knowledge: Forums allow the sharing of ideas; wikis may be used for collaborative group work; databases allow participants to share any type of media, such as photographs or soundtracks.
3. Learning takes place by just observing the activity of our peers. Moodle allows both teachers and learners to view the activities of themselves and others. When a student is able to see that other students have performed and handed in a given assignment, then a student might feel the necessity to perform and submit his/her own work. The visibility of online users shows who from the class is currently online, allowing immediate connection via the Moodle messaging system.
4. Understanding the context of others enables teaching in a more transformational way. The user profile enables the course participants to present information about themselves such as their location, educational and cultural background, or any other personal information they would like to present to the group. The individual blogs that are part of Moodle provide a good opportunity for expressing personal ideas publicly. Forums allow members to share and perceive the minds and opinions of the learners in the group.
5. The flexibility and adaptability of the learning environment offer possibilities to respond to the needs of the course participants.

Moodle is modular, which is very convenient as the components can be added or deleted according to the needs of the group. As well, the content can be activated or deactivated as a need is identified. Some authors, however, express reservations that LMSs are just tools designed alongside the theories of social constructivism and that just using the tools does not ensure the application of constructivist principles—the instructor needs to apply the methods and approaches that go in line with constructivist theories. McLoughlin and Lee (2008) express the views that LMSs just replicate traditional paradigms in online environments, where traditionally students are approached as information consumers. Authors Bryer and Chen (2012) note that LMSs give limited opportunities for online sharing and collaboration as student interaction activities are restricted to one class or one semester while, in comparison, social media tools give a constant opportunity of sharing many-to-many. Although LMSs have social media features, there are additional institutional security and privacy requirements which do not allow sharing beyond the limits of an institution. Mason and Rennie (2008) point out that LMSs should allow students to personalize and customize their own learning experiences rather than aiming to personalize and provide personalized instructions.

> The e-learning environment in LMSs should provide opportunities for students to learn how to: select, combine, coordinate their cognitive strategies in connection to the new knowledge, and be prompted to reflect on their strategies, extending their metacognitive knowledge with strategy and capacity beliefs. Despite the strong recommendation, the LMS is often used as a "one size fits all" service to learners, irrespective of their knowledge level, goals, and interests. All students have access to the same instructional material and the same web-based tools without personalized support. All students receive the same exercises irrespective of their pre-existing knowledge and experience. It is not taken into consideration that educational material is presented to a large number of learners who have varied knowledge level, skills, and learning strategies. (Vovides et al. 2007, p. 68)

The authors stress that an LMS environment should be organized in such a way as to make it flexible and support student self-regulation during learning processes so that it fosters self-directed and self-organized learners who are ready to take an active role in their own learning.

1.3.3 Social Media Choices According to the Learning Objectives

Social media has many varied forms, including Internet forums, blogs, wikis, social networks, podcasts, photographs, video, and social bookmarking. Social networking sites are platforms that allow easy communication and sharing between their members and thereby establishing social networks. Facebook is one of the most popular social networking sites and one of the most researched ones. Blogs are diary-type Web pages where authors write regular entries, and the format of a Web page allows the readers to engage in discussions with the author, with each other, and with other authors. Wikis are Web pages that allow collaborative content creation as the users can contribute and modify the entries (Šliogerienė and Valūnaitė Oleškevičienė 2014).

Short videos allow information sharing through Web pages containing video-sharing services. Podcasts are short MP3 format audio files that could be downloaded onto the computers and then listened to at the time chosen by the users.

Podcast Web pages allow the sharing of audio information. According to Mason and Rennie (2008), photographs are more informative and expressive than words, and they convey the content even in the contexts where language could not be understood or mastered properly. Internet forums are online sites where people could communicate and exchange information in the form of posted messages. Social bookmarking sites are Internet sites that allow people to share and organize online resources and store bookmarks.

If teacher beliefs are based on a social constructivist approach viewing learning as based on social collaboration, the use of wiki media could be a choice for implementing group projects. In onsite environments, students are organized into groups to work together on project tasks, while wiki media allows students to collaborate online fully without meeting face-to-face. Duffy and Bruns (2006) summarize the use of wiki media as follows:

- To develop projects by using wikis for documentation of project work;
- For producing collaborative bibliographies by providing reading resources and short summaries of them;
- To produce concept maps by collaboratively mapping ideas;
- For creating documents collaboratively edited and created by the whole group.

Blogs are different from wikis as they do not have the characteristic of multi-authorship; blogs are usually created and run by one author inviting comments from the blog followers. If the learning objective is to encourage students to express their ideas, to publicize them, and to foster their feeling of being confident and outspoken, then blogs seem to be an appropriate media choice. If the learning focus is to develop students' verbal articulation, then podcasts may seem a reasonable choice. What is more, if an open discussion needs to be encouraged, and if students' negotiating skills, expression of opinions, and exchanging of ideas are the learning objectives, then computer conferencing or Internet forums seem to be an option.

1.4 Social Media and Emerging Educational Approaches in University Studies

Discussions on social-media-generated approaches to university studies are overviewed in this section. Note is taken of emerging educational theories induced by technological change and related to learning-based social media use. It is observed that the initial exaggerated enthusiasm concerning social media use in university studies is replaced by a more sober realistic understanding of social media integration into university study contexts.

1.4.1   Social-Media-Generated Approaches to Teacher–Student Interaction in University Studies

> If you took a doctor from the 19th century and put her in a modern operating theatre, she would have no idea what to do, but if you put a teacher from the 19th century into a modern classroom, she would be able to carry on teaching without a pause. The idea remains that students are empty containers which the teacher fills with the knowledge. (Papert 1993, p. 2)

When reflecting on the above quote from an English textbook, which I used for teaching in 2006, I think that in a modern classroom equipped with all the technology a teacher from the nineteenth century would get puzzled; but, concerning educational approach, the idea of the needs to fill students with knowledge and to criticize unexpected ideas or attitudes, and the desire to instruct, control, test, and measure still prevail in many educational contexts. It is felt, however, that the constructivist approach or student-centered learning is more compatible with Web 2.0 tools and technologies (Laurillard 2002; Beetham and Sharpe 2007).

An open-ended approach, which is gaining influence in university studies, based on constructivist theory, provides students with opportunities to contextualize learning and negotiate knowledge in a collaborative way, which is in line with the basic ideas of constructivism:

- Learning is an active process of constructing knowledge rather than acquiring it.
- Instruction is a process that involves supporting that construction rather than communicating knowledge (Duffy and Cunningham 1996).

The learner-centered model is favored; this supports and guides the learner in the process of constructing the learner's understanding of the reality of which he/she is a part (Duffy and Cunningham 1996; Laurillard 1999, 2002). Researchers also point out the importance of learning based on authentic tasks embedded in the context as learning is viewed as based on sociocultural dialog (Lave and Wenger 1991). In fact, there could be observed an interrelated process whereby technological development provides ICT-rich learning environments including social software tools while at the same time fueling the change of learning paradigms, which is identified by researchers as a need for educational change.

> And requires the development of learning episodes for students that have dialogue and communication as core features. From this perspective, there is a far greater emphasis on networked rather than linear models of learning and on providing culturally relevant experiential and purposeful learning episodes that than the consumption of abstract knowledge.... (Rudd et al. 2006, p. 5)

According to Beetham and Sharpe (2007), a learner-centered approach involves engaging learners in the process of acquiring different kinds of knowledge practice and new processes of inquiry, dialog, and connectivity. The authors identify some key features of learning that stand out as important in the new learner-centered approach:

- Digital competencies that focus on creativity and performance;
- Strategies for metalearning, including learner-designed learning;

- Inductive and creative modes of reasoning and problem solving;
- Learner-driven content creation and collaborative knowledge building;
- Horizontal (peer-to-peer) learning and contribution to communities of learning (e.g., through social tagging, collaborative editing, and peer review).

What is more, the developing and newly emerging social media requires certain skills that its users have to master. Rheingold (2010) identifies five key "literacies" for the effective use of social media:

- Attention, which means the ability to control, direct, focus, or distract one's attention while surfing various types and sources of social media. It is important not to get overwhelmed with details within the social media world but to get one's attention focused on the necessary bits.
- Participation, which means being a good participant who is able to participate appropriately, who knows or has that natural feeling of when, where, and what to post or comment to make the sharing and creation process fluent.
- Collaboration, which is closely related to the collaborative nature of social media where the efforts of many add to the collaborative product creation. The ability to embrace contributions of the members of the collaborative process is of key importance to be able to collaborate in social media or the real world.
- Network awareness, which means mastering the network nature and understanding how networks work, and being able to establish and control one's profile in a network.
- Critical consumption, which means being able to navigate the pool of information and being able to detect meaningful and useful units. Sometimes, Rheingold refers to it as "crap detection." It is important to be able to find reliable core information and use it appropriately.

As social media forms a part of our realities and seems here to stay due to its rapid development, a responsible attitude would be to embrace the reality and equip oneself with the necessary skills—"literacies"—to be able to live and deal with our realities well. Educational theories are thus naturally influenced by social media development, and new concepts and approaches seem to thrive in the environment permeated by social media. Ashton and Newman (2006) break down the broad term "pedagogy," identifying "pedagogy" (teaching of children), "andragogy" (teaching of adults), and "ergonogy" (teaching people to work); they go further in recognizing the need for learner autonomy and self-directed learning by considering heutology, in which learning is totally controlled and guided by the learner (Kenyon and Hase 2010). In a generic sense, the term "pedagogy" is also used to denote "teaching people." The emergence of new pedagogical approaches marks new demands for learning and working in a social media environment characterized by a highly social collaborative nature and constant immersion in the "ocean" of information (Table 1.4).

Laurillard's (2002) conversation approach includes reciprocal changes in an ongoing learner–teacher interaction so that redescription of concepts and actions takes place in a continuous dialog and exchanges. Laurillard (2002) identifies certain basic principles enabling reciprocal learner–teacher interaction:

Table 1.4 New emerging pedagogical approaches

Pedagogical approach and author	Main insights	Pedagogical ideas
Conversation approach (Laurillard 2002)	The conversational approach looks at the ongoing learner–teacher interaction, at the process of negotiation of views of the subject matter which takes place between them in such a way as to modify the learner's perceptions	Students try to learn relationships among the concepts and ideas through explicit conversations with teachers regarding subject matter. Student understanding is facilitated through reciprocal dialogue
Connectivism (Siemens 2004)	An approach that integrates principles explored by chaos, complexity theory, and networking mainly stating that making and sustaining connections are more important than simply knowing	Learning process is characterized by connecting various types of information and by enabling learners to see the connections between concepts and ideas. Learners need competencies what to learn and the meaning of incoming information
Navigationism (Brown 2005)	Navigationism is a more inclusive term than constructivism, and it includes knowledge creation and ability to manipulate, evaluate, and navigate knowledge as well as being able to share knowledge in the process of knowledge creation	Learning is about learner interaction with information and people and about the skills and competencies learners need to survive in the knowledge era

After McLoughlin and Lee (2008)

- The teacher provides intrinsic feedback while controlling the learning environment.
- The learner modifies one's actions in light of feedback and also reflects on learner–teacher interaction to modify redescription.
- The teacher reflects on learner's redescription and action to modify his redescription and the task, and it continues as a reciprocal process.

Laurillard's (2002) approach includes constant reciprocal change; however, Siemens (2004) moves a step forward. Siemens (2004) describes connectivism as: "the integration of principles explored by chaos, network, and complexity and self-organization theories" and presents learning as a process which "is focused on connecting specialized information sets, and the connections that enable us to learn more are more important than our current state of knowing." Siemens (2004) identifies the following essential learning skills and principles:

- Many learners will move into a variety of different, possibly unrelated, fields over the course of their lifetimes.
- Informal learning is a significant aspect of our learning experience. Formal education no longer comprises the majority of our learning. Learning now occurs in a variety of ways—through communities of practice, personal networks, and completion of work-related tasks.
- Learning is a continual process, lasting for a lifetime. Learning- and work-related activities are no longer separate. In many situations, they are the same.
- Technology is altering (rewiring) our brains. The tools we use define and shape our thinking.
- The organization and the individual are both learning organisms. Increased attention to knowledge management highlights the need for a theory that attempts to explain the link between individual and organizational learning.
- Many of the processes previously handled by learning theories (especially in cognitive information processing) can now be off-loaded to, or supported by, technology.
- Know-how and know-what are being supplemented with know-where (the understanding of where to find knowledge needed).

Regarding navigationism, Brown (2006) focuses on student ability to navigate the surplus of knowledge; he states that it is important for students to be able to collaboratively explore, evaluate, manipulate, and integrate knowledge available in various sources and modes. Brown (2006) identifies certain skills and competencies required in a navigationist paradigm:

- The ability—know-how and know-where—to find relevant and up-to-date information, as well as the skills required to contribute meaningfully to the knowledge production process. This includes the mastery of networking skills and skills required to be part of and contribute meaningfully to communities of practice and communities of learning. This implies that the basic communication, negotiation, and social skills should be in place;
- The ability to identify, analyze, synthesize, and evaluate connections and patterns;
- The ability to contextualize and integrate information across different forms of information;
- The ability to reconfigure, represent, and communicate information;
- The ability to manage information (identify, analyze, organize, classify, assess, evaluate, etc.);
- The ability to distinguish between meaningful and irrelevant information for the specific task at hand or problem to be solved;
- The ability to distinguish between valid alternate views and fundamentally flawed information. Sense making and chaos management (Brown 2006).

As well, Brown (2006) provides a summary of the connectivist learning skills and principles required within a navigationist learning paradigm:

- Learning is a process of connecting specialized nodes or information sources.
- The capacity to know more is more critical than what is currently known.

- Nurturing and maintaining connections are needed to facilitate continual learning.
- The ability to see connections between fields, ideas, and concepts is a core skill.
- Currency (accurate, up-to-date knowledge) is the intent of all connectivist learning activities.
- Decision making is itself a learning process. Choosing what to learn and the meaning of incoming information are seen through the lens of a shifting reality. While there is a right answer now, it may be wrong tomorrow due to alterations in the information climate affecting the decision (Brown 2006).

If we compare the lists, it could be seen that certain learning principles intertwine in both theories. The emergence of new theories and the need for them are related to the nature of social media, which turns learning from an act of an individual nature into an act of a collective nature embedded into the new realities of social media. If we still remember Oakeshott (1989) and his holistic approach to the essence and function of university education, which is "teaching to see," it seems that a holistic approach to learning and living in any reality has been embedded already in holistic approaches by Lao Tzu: "Earthly truths are limited and contradict each other. They lead to common truths but common truths do not obey order and lead to what cannot be ordered" (Lao Tzu 2009, p. 54).

The above becomes so visible now with the rapidly developing technologies creating new emerging realities and the need for a learner to be ready to constantly navigate and learn in the fluid world of ever-changing information and technology. What existed yesterday has been changed by today's reality, what exists today is going to be changed by tomorrow's reality, and people need to learn to live in the constant change.

1.4.2 Flexible Learning

It appears that student-centeredness, flexibility, interactivity, and a dynamic learning environment are the central features of a rationale for the choice of media and methods. Mason and Rennie (2008) point out that flexible learning is not a new phenomenon. In fact, learning takes place in both the environments of formal education and out of formal settings. Collins and Moonen (2001) defined four key features of flexible learning:

- Technology;
- Educational approach;
- Implementation strategies;
- Institutional framework.

Most importantly, though, they stressed the necessity for flexible learning to be focused on the benefits of the learner.

> Flexible learning is a movement away from a situation in which key decisions about learning dimensions are made in advance by the instructor or institution, towards a situation where the learner has a range of options from which to choose with respect to these key dimensions. (Collins and Moonen 2001, p. 10)

Notwithstanding this, Biesta (2013) tries to convey the view that the abundant use of notions such as "learning" and "learner," which he even denotes as the process of "learnification," should not remove our focus from the importance of relationships in educational processes and the importance of the responsibilities and tasks of the educators. In fact, the author questions the shift from teaching to learning as a naturally contradictory move since too much stress on learning as an individual and individualistic act moves the focus from pedagogy toward a learning theory and diminishes the role of the teacher and educational approaches. The application of new technology such as social media largely depends on the attitude of the educational institution as a whole unit. As Thompson (2007) states,

> Web 2.0 is a potentially disruptive technology because of its potential to change the model of university studies from traditional classroom framework to asynchronous 24/7 mode. Institutes of higher education historically do not cope well with disruption, especially in short term; however, coping with this disruptive force could mean engaging students in extended collaborative learning opportunities. From this perspective, the perceived disruption could entail many positive implications for higher education. (Thompson 2007, p. 9)

Thompson sounds both reserved and optimistic at the same time as the author acknowledges that institutions of higher education—including universities—need time for adoption of new technologies but that there are also extended learning opportunities within the implementation and use of the technology. A similar insight is expressed by Lamb:

> Educators and higher education decision-makers have an obligation to carefully and critically assess new technologies before making radical changes. Taking a more freewheeling approach to content reuse and making campus technologies more accessible to data mash-ups require significant changes in existing practices and attitudes. These changes won't happen quickly and easily. (Lamb 2007, p. 13)

The educator concerns are related to the following main issues:

- The academic staff may feel that they need additional training for the use of new technologies and new technological applications.
- The academic staff have to move toward more student-directed learning.

Tapscott (1998) identified interactivity of learning as an educational technique that requires change in education and listed the areas of change as:

- From linear to hypermedia learning;
- From instruction to construction and discovery;
- From teacher-centered to learner-centered education;
- From absorbing material to learning how to navigate and how to learn;
- From schooling to lifelong learning;
- From one-size-fits-all to customized learning;

- From learning as torture to learning as fun;
- From the teacher as transmitter to the teacher as facilitator.

In fact, social media use has a tendency to shift focus from a teacher to a facilitator of learning, which does not mean that a teacher ceases to teach; it just means that the pedagogical focus shifts and the pedagogical relationship shifts toward a more democratic one. Veen and Vrakking (2006) also admit that learning is not just a one-way process and that technology can be used to enhance interactive learning.

It is not as straightforwardly easy as it may seem at first sight, though. Social media just used by itself will not make the existing face-to-face course more interactive, nor will it solve the problems of the existing course. If the onsite course material, handouts, and reading resources are directly uploaded onto social media sites without any modifications, the problems like lack of student participation remain the same; to reformulate the thought, all the advantages and disadvantages of the existing onsite courses are pasted onto social media Web sites. In this way, social media does not work as a panacea. In fact, an educational approach—the understanding of the educational aim—remains of key importance (Biesta 2013). The choice of media and activities within it depend on the educational goal. Social media could be used in a variety of ways, and what works in one situation does not necessarily work in another one. The responsibility of a teacher is to orchestrate the technological means and course materials in tune with the educational goal.

What is more, it is important to make sure the students are comfortable with the technology as while using social media students should focus attention on the activity instead of wasting their time on figuring out social media applications or being distracted by feeling uncomfortable with the technology.

It is also important that activities are worthwhile, related to real-life experience, and have clear learning benefits as students are concerned about using their time effectively. In an attention economics approach, human attention is treated as a scarce commodity. According to Davenport and Beck (2001), attention focused mental engagement on a particular item of information. Items come into our awareness, we attend to a particular item, and then we decide whether to act. Within the growing abundance of information in social media, human attention becomes a limiting factor such that clear and transparent learning benefits might become a factor for retaining students' attention; in the opposite situation, students may focus their attention and spend time elsewhere.

Chapter 2
Research Methodology

2.1 Choice of Research Methodology

Historically, educational research has undergone changes, moving from a quantitative approach toward a qualitative approach which allows researchers to successfully apply both approaches as complementing each other. The criticisms of the application of purely quantitative scientific methods in educational research include such arguments as the fact that human social life cannot be simply characterized by a mechanistic cause-and-result understanding, and that human interaction involves complex processes of negotiation and interpretation and at times determined outcomes cannot be set (Creswell 2007). The author argues that qualitative research is applicable in researching educational processes that are essentially characteristic of ongoing interpretation and interaction. Otherwise, scientific methods used in quantitative educational research may lead to practices, which might result in standardization (Duoblienė 2011).

Qualitative research applies methods that allow the researcher to capture stories of participants' own experiences and allows them to achieve the meaning of experience. For this study of social media use in university studies, I have chosen inductive qualitative research with a phenomenological approach as the research is intended to reveal how research participants make sense of their experiences related to social media use in studies. The meaning that research participants give to shared lived experience helps in examining how things really are and what scientific insights could be applicable. Although theoretically it could be known how things should be, education in particular is a sensitive area where regulations and instruction may clash with human realities.

Thomas (2006) observes that there are many specific approaches to analyze qualitative data, such as grounded theory (Strauss and Corbin 1990), phenomenology (e.g., van Manen 2014) or narrative analysis (e.g., Leiblich 1998); however, there are more generic approaches to qualitative data analysis that represent inductive qualita-

G. Valunaite Oleskeviciene and J. Sliogeriene, *Social Media Use
in University Studies*, Numanities - Arts and Humanities in Progress 13,
https://doi.org/10.1007/978-3-030-37727-4_2

tive research (e.g., Silverman 2005; Ezzy 2002; Elo and Kyngäs 2007). The inductive qualitative approach is a systematic procedure applied in order to analyze qualitative data (Silverman 2005; Elo and Kyngäs 2007) with the essential purpose of allowing research findings based on dominant and significant themes to emerge from the raw data. Such an approach allows the identification of essential meanings of the research participant experience.

Additionally, using a phenomenological approach enables us to see the real state of the things (Saevi 2012). At this point, I always remember a comparison given by a former professor of mine: Engineering researchers may throw cars at the wall aiming to check car endurance, but it is impossible to throw humans against the wall to check human endurance. A phenomenological approach allows us to sensitively register human realities in education. Social media in university studies is used both internationally and nationally, acting as a driving force enhancing technology-driven processes. The choice of a phenomenological approach was inspired by the complexity of the discourses concerning the phenomenon of social media use in university studies and also by the nature of the phenomenon based on the experience of people: teachers, students, and administrators who actually apply social media in university studies, who actually experience the phenomenon of our living world. Phenomenological research is based on the approach that every person creates their own personal educational reality based on personal experience. Additional reasons for choosing this approach embrace the fact that a phenomenological approach acknowledges the relative subjectivity of the researcher and the relativity of the research; this does not lead to generalizations but rather leads more to a sensitive understanding of human conditions in education and recognizing the value of research participant voices on the topic.

2.2 Inductive Qualitative Research Procedures: Data Collection and Data Analysis

In terms of inductive qualitative research procedures, the following main steps can be distinguished (Thomas 2006; Creswell 2007; Elo and Kyngäs 2007):

• The research phenomenon and aim are identified;
• Empirical material is collected;
• The data is analyzed;
• The researcher constructs a description revealing the essential aspects of the most important themes in the raw data.

Another important issue is the research sample. The number of participants in the sample depends on the researcher's commitment level to the analysis of cases (deeper-less-more); however, the main factor is the saturation, when the researcher notices that new details no longer appear and things become repetitive. Also, it is important for the sample size to represent the richness of the individual cases and whether the researcher compares or contrasts the experiences. In the selection of the study participants, the ability of participants to reflect on their experiences must be taken into account. There are some pragmatic difficulties a researcher may encounter. First, it is advisable to obtain a more homogeneous sample. Researchers usually try to create homogenous samples—that is, to choose the people for whom the research issue would be significant. The decision making is sometimes a practical problem (Which people is the situation relevant to? How easy is it to make contact with them?). Additionally, there is the other partially interpretational problem: Are these people different from each other, and how many of these differences are permissible within a given study (Creswell 2007)? The other issue is the possibility of obtaining the participant's background information as it is not always openly available.

In terms of data collection, researchers (Smith et al. 2009) prefer to enter the research field already having prepared a semi-structured interview, what they call a "schedule" which helps to facilitate an interaction with a research participant. It is important for the researcher to apply good tactics for conducting interviews, which include the following (Smith et al. 2009):

• Asking sensitive questions later in the interview process;
• Selecting the pace and rhythm of the conversation;
• Teaching the participant to engage in a conversation;
• The importance of clarification for the emergence of details.

The main methods of data collection can be identified as: semi-structured interviews, unstructured interviews, diaries, and autobiographical narratives (Elo and Kyngäs 2007).

Open-ended questions are used in interviews to get rich and detailed descriptions of the phenomenon being studied. Such questions might include "What is the feeling of being unemployed?" Research questions focus on the important aspects of the experience at critical moments or everyday life. They may include (Creswell 2007): Hot cognition, which involves current problems or dilemmatic situations, or cold cognition, which deals with the long-term life-experiences reflection.

In terms of data analysis, the researchers Smith et al. (2009) acknowledge both manual data processing and using qualitative data software as equally valid. According to the authors (Elo and Kyngäs 2007) the main three phases can be identified. First, the interview content is transcribed. Then, the transcribed interviews are read a lot of times while writing notes in the right margin about generalizations, associations, relations, and refining the concrete experience; in the left margin notes are written regarding listing emerging sub-theme titles—they may be imperfect, but they should help to articulate what is thought about or what could be summed up. After that, sub-themes are combined into clusters of themes, which make super-ordinate themes. Concerning the notion of super-ordinate themes or categories, some authors (Thomas 2006; Elo and Kyngäs 2007) use the notions "super-ordinate themes" and "categories" interchangeably while others choose one definite notion.

Doing interpretative work with data, repeating themes and important themes are highlighted. The following main parts of the theme formation process could be distinguished (Thomas 2006; Elo and Kyngäs 2007):

- The text is organized into themes;
- Themes are supported by certain extracts;
- Themes are merged into clusters, according to the mutual relationships;
- Clusters can be connected to the super-ordinate themes;
- A table of all the clusters and themes is organized;
- Each theme is shown to be an accurate transcription quote, without any alterations by researcher.

Detailed descriptions and quotations to illustrate the meanings of the developed themes are used in reporting inductive qualitative findings. The way in which all the cases were analyzed is the following: The researcher starts from the case 1 themes and looks for the evidence of this theme with case 2. New themes may appear in the experience of case 2. As it is a cyclic process, the researcher returns to case 1 and quickly reviews whether this new theme is also in case 1. This way the researcher goes through the steps of the analysis several times, comparing cases with each other changing and adjusting themes and clusters. A subjective creative work is produced, but inductive qualitative research does not require purified objectivity. Another investigator may find some different nuances (Smith et al. 2009).

While discussing the data, it is good to summarize taking into account a number of different contexts; then, the logical connections are set and a clear picture of the phenomenon interpretation from various positions of participants is created. Individual experiences are used for justification, but they are also abstracted. The research findings are interpreted as hypothetical. They should be checked further, but for this, another sample should be collected. Inductive qualitative study has limited generalization possibilities.

Data discussion is the phase that results in placing theoretical paradigms or insights (interpretations) into a particular context. We can discuss and compare what the collected data reveals and what is claimed according to a particular theory. This interpretation is additional, of a secondary character. It provides the text and experience with even more colors, and it reveals a secondary sense, conflicts, and resistance—what the research participant does not see.

As for the validity of the analysis, the data are presented and discussed with experts, they can sometimes be discussed with the participants themselves, and they are confirmed by the quotes from the same dataset (Creswell 2007). A clear description of the procedure is also important. Triangulation can be used at any level; for example, using not only the interview but also the diary method, interviewing several groups of research participants providing the experience (parents–children) or using a number of experts. The final word is the researcher's—it is his/her decision, depending on the research aim.

The researcher has to decide which themes focus on the research question and select citations. The researcher's choice does not necessarily depend on the frequency of occurrence of the theme; it also depends on the richness of a particular experience revealing the theme and how the theme illustrates other important issues. When planning the time, one needs to keep in mind that the analysis is time-consuming, and it is necessary to allocate a lot of time while designing the research.

2.3 Phenomenological Approach

The etymological meaning of "phenomenology" could be broken into the meanings of the two words: "phenomenon," which means what could be seen, or, according to the *Merriam-Webster Online Dictionary*, "something (such as an interesting fact or event) that can be observed and studied and that typically is unusual or difficult to understand or explain fully"; and "logos," which means speech or study or, in the *Merriam-Webster Online Dictionary*, "the divine wisdom manifest in the creation, government, and redemption of the world" or "reason that in ancient Greek philosophy is the controlling principle in the universe." We can see that "logos" have the meaning of divine wisdom or reason, the controlling principle of the universe, which reveals that "logos" embodies the ability or the means to explain, to reveal the wisdom or the controlling principle of what is observed.

According to Heidegger (1972), "Phenomenology means: To let what shows itself be seen from itself, just as it shows itself from itself." That is the formal meaning of the type of research that calls itself "phenomenology." But it expresses nothing other than the maxim formulated above: "To the things themselves!"

What is important here is that the aim of phenomenology is to expose the essence—the meaning that is in the phenomenon itself—and how the phenomenon reveals itself in the course of experience, how it is experienced. Gadamer (1999), analyzing the two sources of modern thinking—logos and mythos—reveals that one source in essence is enlightenment ideas on which Anglo-Saxon tradition, modern science and technologies are based; the other one is mythos, which is a story that is told, and it cannot be experienced in any other way than by hearing it. They both are intertwined, helping human "Dasein" express and understand what shows itself. In addition, understanding is based on the hermeneutic circle, where the whole is perceived based on separate cases and separate cases are perceived based on the whole.

Phenomenology involves not only the methodology but also philosophy. It is based on the works of Edmund Husserl and his followers, such philosophers as Heidegger, Sartre, and Merleau-Ponty (van Manen 1990). Phenomenological philosophy emerged in response to the scientism that prevailed at the end of the nineteenth century—the philosophical attitude stating that an objective knowledge of reality (and even social reality) can only be based on the natural sciences and their methodology. Phenomenology highlighted the peculiarity of social phenomena and sought to distance itself from the prior understanding of a "real social phenomenon" in order to explain the phenomenon as a conscious experience of individuals, stating that a social phenomenon is what is experienced by a variety of individuals.

Phenomenology examines the human experience since the phenomenon is known to people through their personal experiences. This approach is particularly interested in social existence; it stems from a philosophical interest in the essence of being-in-the-world (Heidegger 1962). Phenomenology has become popular in the social sciences, especially sociology, psychology, and educational science. Phenomenology is sometimes presented as an alternative to positivism, as this approach emphasizes:

- Subjectivity (versus objectivity);
- A description (versus analysis);
- Interpretation (versus measurement);
- Factors (versus structure).

The contrast with positivism is strengthened further by the fact that phenomenology examines the lived experiences of people based on the claim that these experiences are conscious. The phenomenological researcher tries to describe the nature and essence of these experiences rather than to analyze or explain them.

Currently, the variety of phenomenological methodologies and multiplicity of methods and issues analyzed are observed (Embree 2001; Gill 2014). The phenomenological tradition embraces a variety of tendencies within multiple schools of phenomenology and multiple disciplines and sub-disciplines. Among the developing variety of multiple phenomenological methodologies, Gill (2014) presents and discusses five phenomenological methodologies by comparing the varying assumptions, aims, and steps of analysis. The author observes that both Sander's phenomenology and Giorgi's descriptive phenomenological method belong to descriptive phenomenology, and bracketing is essential for their phenomenological inquiry, while van Manen's hermeneutic phenomenology, Benner's interpretative phenomenology, and Smith's interpretative phenomenological analysis reject the idea of bracketing by acknowledging the researcher's limitations and involuntary presuppositions. The author points out that different methodologies put emphasis on different analytical activities; for example, van Manen's hermeneutic phenomenology is focused on depthful writing aiming to transform lived experience into a textual expression of its essence. Other researchers focus more on identifying meaning units and possible themes in order to obtain the structure of the experience. Some phenomenologists (e.g., Lindseth and Norberg 2004) suggest structurally condensing the meaning units found in the raw data into sub-themes and themes by using tables. It is evident that various phenomenological methodologies apply a variety of methods of analysis; however, they all embrace the main aim of investigating the meaning of lived experience.

Bearing in mind the multiple phenomenological methodologies, Embree (2001) observes that "there is a great deal of national and disciplinary myopia within phenomenology, which is to say that scholars tend to focus upon what is happening in their own disciplines and schools" (Embree, 2001, p. 2) rather than taking a wider-embracing view. The author expresses his preference for the synthesis of various methodologies and practicing a generic approach.

According to Embree (2001), a phenomenological study aims to describe and understand how different individuals experience a phenomenon and what meanings they attach to the phenomenon. The researcher's focus is on the phenomenon and how the phenomenon is experienced by different individuals. The researcher is looking for what is common to the different experiences of individuals, what all individuals experience faced with the phenomenon (Embree 2001). The common experience, rather than the prior knowledge of the researcher of the phenomenon as a part of the reality, is the essence of the phenomenon. The researcher collects data from individuals who experienced the phenomenon and examines what and how they have experienced. Van Manen (2014) suggests using existentials as universal "themes" to explore the meaning-making of the lived experience; the author introduces such existentials as lived relations (relationality), lived space (spatiality), and lived time (temporality). The fundamental existentials are recurrently applied in phenomenological writing as they guide the researcher in exploring the meaning. The existentials are also employed for an enriched search of the meaning in the current research.

2.4 Research Procedures

An international inductive qualitative study served as a starting point of the research that captured the initial insights on social media use in university studies. Then, a study at home institutions was carried out to gain a deep understanding of the phenomenon of social media use in university studies.

2.4.1 Research Procedures of the International Study

The ongoing advance of social media into university studies has a global character and affects the local processes in university studies driving toward globalization. In fact, global and local get intertwined into oneness. The starting point of the study was international inductive qualitative research. The research was carried out within the framework of the Grundtvig multilateral partnership project "Institutional Strategies Targeting the Uptake of Social Networking in Adult Education (ISTUS)" (De Angelis et al. 2013). From the practical point of view, it was an appropriate starting point for obtaining the initial insights on the phenomenon of social media use in university studies.

Qualitative semi-structured interviews were designed and organized as the way of collecting empirical research data. The reasons were mainly determined by the nature and the aim of the research itself. The research strategies of semi-structured interviews were chosen because the project team aimed to research how university study participants—teachers, students, and administrators—make sense of social media use in university studies through their own lived experiences. Open-ended questions allowed the research participants to expand on their experiences of social media use in university studies.

Following the insights by Ricoeur (2000), communication is crucial in human existence because communication is the main tool that enables sharing of the meaning of lived experience. The experience is always personal, but with communication the meaning of the experience could be shared and becomes public. In this way, communication enables us to overcome the untransferable nature of the lived experience. Ricoeur (2000) shows the capacity of communication to allow transference of the meaning of the lived experienced to others. That is why semi-structured interviews acquire the power to receive the empirical data of the lived experience from the research participants.

Three research participant groups were established: teachers, students, and administrators, in order to represent the phenomenon of the use of social media in education in a broader context, and also to ensure data source triangulation. Interviews were conducted in Germany, Switzerland, Italy, Finland, Lithuania, Netherlands, and UK at different educational institutions that largely consist of universities (5) and one University of the Third Age, which are in our focus of research on the phenomenon of social media use in university studies. The researched universities provide study programs ranging from undergraduate to PhD where learning takes place onsite or in a mixed mode. In the course of the survey, 10 interviews with administrators, 20 interviews with teachers, and 20 interviews with students were conducted. The sample of the research participants is presented in Table 2.1.

The method of criterion-based sampling was applied. Respondents had been selected according to the following criteria: nature of institution (specialists who work or students who study at universities) and experience using social media (administrators and lecturers with at least three years of job experience using social media). Also, the research team agreed to consult the offices responsible for academic and scientific affairs in the institutions before planning the interviews to get their recommendations for choosing research participants who are known as successfully using social media for education purposes. Time for each interview was personally allocated, having been discussed with each interview participant. At the beginning of each interview, the aim of the research was presented to the interviewees. Also, their active participation, objectivity, and sincerity were encouraged during the interviews. Research participant consent to participate in the interviews was received from the participants, and the researchers reassured the participants of the confidentiality of the data. The interviews were carried out in the native languages of the research participants and then the data was translated into English.

During the interviews, open questions were used to get the data on the explicit nature of the use of social media either for work or studies. The following aspects were discussed during the interviews:

- Background of the institution;
- Social media use in research participants' daily jobs or study routines;
- Influence of social media on work and learning environments experienced by the research participants;
- The most memorable experience of the research participants concerning social media in university studies;
- Institutional influence experienced by the research participants while using social media in teaching and learning.

Finally, the last step was the analysis of the interview data that was based on inductive qualitative analysis, which included the following parts:

Table 2.1 Research participants of the international study

Teachers

Research participant	Position	Work experience at university	Level of study programs taught	Country
T1	Senior lecturer	10	Master's	Finland
T2	Senior lecturer	12	Master's	Finland
T3	Lecturer	9	Bachelor's	Finland
T4	Lecturer	7	Bachelor's	Netherlands
T5	Lecturer	10	Bachelor's	Netherlands
T6	Lecturer	5	Bachelor's	Netherlands
T7	Lecturer	10	Bachelor's	Switzerland
T8	Lecturer	9	Bachelor's	Switzerland
T9	Associate professor	15	Master's	Switzerland
T10	Senior lecturer	10	Master's	Italy
T11	Senior lecturer	7	Master's	Italy
T12	Senior lecturer	10	Master's	UK
T13	Senior lecturer	10	Master's	UK
T14	Lecturer	5	Bachelor's	UK
T15	Lecturer	4	Master's	Germany
T16	Lecturer	7	Master's	Germany
T17	Associate professor	15	Master's	Germany
T18	Associate professor	15	Bachelor's degree	Lithuania
T19	Lecturer	7	Bachelor's degree	Lithuania
T20	Lecturer	9	Bachelor's degree	Lithuania

Administrators

Research participant	Position/Job nature	Work experience at university	Supervised study programs	Country
A1	Business unit manager	10	Bachelor's and Master's	Netherlands
A2	Marketing, product development	9	Bachelor's and Master's	Netherlands
A3	Director of one of the institutes	10	Bachelor's and Master's	Switzerland
A4	Academic development advisor, staff development	20	Bachelor's, Master's, and Doctoral	UK

(continued)

Table 2.1 (continued)

Administrators

Research participant	Position/Job nature	Work experience at university	Supervised study programs	Country
A5	Faculty dean	10	Bachelor's and Master's	UK
A6	Web administrator, technical support for teachers	10	Bachelor's, Master's, and Doctoral	Italy
A7	Administrator at university	15	Bachelor's, Master's, and Doctoral	Italy
A8	Head of managing board	7	Master's	Germany
A9	Vice chancellor	10	Master's	Germany
A10	E-studies group member	12	Bachelor's and Master's degree programs and doctoral studies	Lithuania

Students

Research participant	Study year	Study level	Employed	Country
S1	3	Master's degree	Yes	Germany
S2	2	Master's degree	Yes	Germany
S3	1	Master's degree	Yes	Germany
S4	1	Master's degree	Yes	Finland
S5	1	Master's degree	No	Finland
S6	2	Master's degree	Yes	Finland
S7	1	Master's degree	Yes	UK
S8	3	Bachelor's degree	Yes	UK
S9	2	Bachelor's degree	No	UK
S10	3	Master's degree	Yes	Switzerland
S11	3	Bachelor's degree	Yes	Switzerland
S12	1	Master's degree	Yes	Netherlands
S13	2	Master's degree	Yes	Netherlands
S14	2	Bachelor's degree	No	Netherlands
S15	2	Master's degree	Yes	Italy
S16	2	Bachelor's degree	No	Italy
S17	1	Bachelor's degree	No	Italy
S18	3	Bachelor's degree	Yes	Lithuania
S19	2	Bachelor's degree	Yes	Lithuania
S20	1	Bachelor's degree	No	Lithuania

- Identification of key aspects represented by phrases and parts of the sentences, allocation of keywords to represent the arising themes, and supporting themes with certain extracts;
- Merging of themes into clusters of meaning according to the mutual relationships;
- Grouping of clusters into the super-ordinate themes;
- Providing descriptions of the developed themes.

The themes and super-ordinate themes were discussed and validated by the international researcher team.

2.4.2 Research Procedures of the Inductive Qualitative Study with a Phenomenological Approach at Home Institutions

In home institutions, the same three research participant groups were identified as in the international study—teachers, students, and administrators—to ensure data triangulation, and the method of criterion-based sampling was applied. Respondents had been selected according to the following criteria: nature of work (specialists who apply social media in their work with students) and experience using social media (administrators and teachers with at least three years of job experience using social media). The group of teachers consists of eight research participants and the group of students consists of nine research participants. The group of administrators includes four research participants and consists of the employees directly responsible for e-learning, various electronic resources, use of technologies, and social media in studies. The administrators perform both technical and administrative work and make up a digital studies unit managing technology and social media adoption at the university. In addition, to ensure the homogeneity of the sample, teachers from the same departments who teach foreign languages were chosen, and the student respondents were chosen according to the recommendations of the teacher respondents who teach them and who could recommend them as active social media users who could provide rich insights on technology and social media use in university studies. The sample of the study at home institution is presented in Table 2.2.

Time for each interview was personally appointed, having been discussed with each participant of the interview. At the beginning of each interview, the aim of the research was presented to the interviewee, and their active participation, objectivity, and sincerity were welcomed during the interview. Permission to record the interview was received from each participant, and the researchers reassured the participants of the confidentiality of the data. The number of interviews was decided according to the saturation principle; the interviews were ceased when new themes stopped appearing. The research participants were questioned on their experiences of using social media either for work or studies, on how they use social media for their work or studies, and on the most memorable experience they could account for. A brief description of each research participant is provided to present their backgrounds and

Table 2.2 Research participants of the study at home institution

Teachers

Research participant	Position	Work experience at university	Level of study programs taught
T1	Associate professor	25	Bachelor's degree
T2	Lecturer	9	Bachelor's degree
T3	Associate professor	19	Bachelor's degree
T4	Lecturer	15	Bachelor's degree
T5	Professor	17	Bachelor's degree
T6	Lecturer	10	Bachelor's degree
T7	Lecturer	7	Bachelor's degree
T8	Lecturer	5	Bachelor's degree

Administrators

Research participant	Position/Job nature	Work experience at university	Supervised study programs
A1	Supervision of e-studies	15	Bachelor's and Master's degree programs
A2	E-studies group member	10	Bachelor's and Master's degree programs and doctoral studies
A3	E-studies group member	5	Bachelor's and Master's degree programs
A4	Academic affairs center	10	Bachelor's and Master's degree programs and doctoral studies

Students

Research participant	Study year	Study level	Employed
S1	2	Bachelor's degree	Yes
S2	2	Bachelor's degree	Yes
S3	1	Bachelor's degree (second degree)	No
S4	1	Bachelor's degree	No
S5	1	Bachelor's degree (second degree)	No
S6	1	Bachelor's degree	No
S7	1	Bachelor's degree	Yes
S8	2	Bachelor's degree	Yes
S9	2	Bachelor's degree	No

perspectives related to their experiences of social media use in university studies. Each participant was given a pseudonym to ensure confidentiality and protect their identity.

In qualitative research, in a way, the researcher becomes an instrument for both obtaining and analyzing empirical data. Considering the interviewer's crucial role in obtaining data, Kvale (1996) provides a metaphor of an interviewer as a traveler who registers what he hears and sees during his travel. The metaphor of travel embodies a double meaning: on one hand, travel is directed toward obtaining knowledge; on the other hand, travel changes the interviewer by stimulating reflection, which leads to new ways of self-reflection and noticing different matters. Thomas (2006), while discussing general inductive data analysis, indicates that data analysis is determined by findings arising directly from the raw data; however, the key framework, in which the central themes are organized, is judged important by the researcher. The metaphor of travel worked for me as a researcher in the process of raw data analysis. It reminds me of the example of physical matter research given by van Manen (2014). Everyone knows that the physical matter, the things we can touch and recognize as solid, in the deep sense have the inherent wave motion nature. The deeper science penetrates into the nature of physical matter the more it reveals how particles and waves dissolve into one another. In fact, they both exist in the constant intertwining change of quantum mechanics. Similarly, when the raw data is analyzed there is an intertwining play of various approaches familiar to the researcher. In this particular case, the process of analysis could be divided into three intertwining layers:

- The first layers of the thematic organization of the phenomenon rely on the two main trends expressed in literature analyzing social media use in higher education—the optimistic approach stressing the advantages and the less optimistic one presenting and analyzing the issues;
- The following layer is deciphering the existentials of lived experience indicated by van Manen (2014)—relationality, spatiality, and temporality;
- Finally, at the deepest level of the analysis, the two key themes stand out—lecturer creativity expression while using social media in university study processes and the pedagogical relationship between lecturers and students while using social media.

In fact, the myth of Orpheus and Eurydice so widely analyzed by phenomenologists (Balnchot 1981; van Manen 2014) symbolically represents the phenomenological way of looking for the truth for the essence of the phenomenon. Orpheus turns around to see Eurydice in the darkness of the underworld; he has a desire to transcend the limits of the living world and get a glimpse of the primordial essence of Eurydice still in the world of death and darkness just before entering the living world again, and Eurydice is momentarily snatched away before he can touch or hug her in the different realm. Similarly, phenomenologists look for the essence of the phenomenon, but it is always only possible to get a glimpse of the truth behind our living world, and still phenomenology gives the researcher an opportunity to get that glimpse.

What I learned as a researcher is that it is hardly possible to achieve one final truth (veritas)—what could be done is approximate trying to get deeper. There are layers of truths, and all of them are correct. If we look at the first layer of advantages and disadvantages of social media use, it could be called opinions but sometimes they are practically useful in making an informed picture on social media use. Looking at existentials helps us to define more clearly our existence with human extensions, as McLuhan (2003) calls them. Finally, getting to the core of the study process we find out how important the pedagogical relationship is for the participants of educational processes, be they in social media or not, and how important lecturers' human creativity stands out in the process of equipping themselves with newest technologies, human extensions, and social media.

Chapter 3
Findings of the Research on Social Media Use in University Studies

3.1 International Study

The findings of the international inductive qualitative research include the descriptions of the emerging themes and their clustering into super-ordinate themes. The analysis allows us to establish major contradictions and positive effects of social media use in university studies.

3.1.1 Findings of the International Study

The analysis of the interviews allowed us to establish the meaning units in the interview material of the research participants. The meaning units were then joined together to form condensed sub-themes, which later were linked or clustered together to form the themes and super-ordinate themes. The whole process with all its steps seeking to define super-ordinate themes is presented below in the tables. The tables are organized to present the three groups of research participants: lecturers, students, and administrators, focusing on social media application in teaching/learning experiences for each research participant group accordingly. Four common super-ordinate themes among the three groups of participants have been distinguished: social media application, perceived advantages, perceived problems, and change manifestations.

3.1.1.1 Teacher Experiences

Introducing the first super-ordinate theme—social media application—teacher use of social media is presented in Table 3.1. It reveals that teachers mainly identify two purposes of their social media use:

G. Valunaite Oleskeviciene and J. Sliogeriene, *Social Media Use in University Studies*, Numanities - Arts and Humanities in Progress 13,
https://doi.org/10.1007/978-3-030-37727-4_3

Table 3.1 Social media application (teacher use of social media)

Meaning unit	Abstracted unit	Sub-theme 1	Sub-theme 2	Theme
I use social media for communication with students and other parallel organizations	Communication with students and organizations	Communication with students	Communication purposes	Teacher use of social media
Announcements to students, answering students' questions, keep students informed	Announcements to students, answer questions			
Communicate with friends, stay updated on friends' and colleagues' news and events	Communicate with friends and colleagues	Communication with friends		
Ask friends for advice	Ask for friends' advice			
To share material, social media is the best way for me to store visual notes online	To share material, to store visual notes	Sharing material	Sharing information purposes	
In my teaching process I usually use Facebook groups to communicate and share materials with my students	Use Facebook groups to communicate and share materials with my students			

- Communication,
- Sharing information.

Teachers point out that social media is a means for communication with students, institutions, colleagues, and friends: "I use social media for communication with students and organizations; with friends and colleagues." The other purpose identified is sharing information, which includes storing teacher notes and sharing the material with students: "To store visual notes, share materials with my students."

Speaking about the second super-ordinate theme—perceived advantages of social media—these teacher insights are shown in Table 3.2. The theme of teacher-perceived advantages of social media embraces numerous sub-themes:

- Promotion of information exchange;
- Increased student involvement;
- Promotion of institution;
- Improved environment.

The sub-theme "promotion of information exchange" is closely related to the use of social media. It includes such factors as abundance of information and opportunity

Table 3.2 Perceived advantages (teachers)

Meaning unit	Abstracted unit	Sub-theme 1	Sub-theme 2	Theme
Social media such as Wikipedia provides useful information	Social media provides useful information	Useful information	Promotion of information exchange	Perceived advantages of social media
Social network is useful to be aware of others' activities	It is useful to be aware of others' activities	Awareness of ideas and activities		
Getting up to date about what other people think	Knowing what other people think			
Reaching more students and communicating in a better way	Reaching more students	More students involved	Increased student involvement	
Promote discussions, improve collaboration among the students	Promote discussions among the students	Promotion of discussions		
Ways to gain feedback increase	Gaining feedback increases	Increased feedback		
Possibilities of selling courses abroad	Selling courses abroad	Selling services	Promotion of institution	
Disseminate university's activity to others. Improve the visibility of the university	Disseminate school's activity	Dissemination		
Social media improves the external image of a university/institution	Improves the external image of a university/institution	Improving image of institution		
Using paper less is good for ecology	Less paper is good for ecology	Better ecology	Economical use	
Networking all over the world, less travel time	Networking all over the world saves travel time	Saved travel time		

to be informed about the activities and ideas of others: "Useful to be aware of others' activities, knowing what other people think," and it identifies the possibility of sharing ideas and various information.

The sub-theme of increased student involvement demonstrates positive teacher outlook on how media could enhance student participation by reaching vast numbers of students and by promoting discussions and feedback: "Promote discussions among the students, gaining feedback increases." Also, teachers identify institution promotion by indicating that social media provides possibilities for disseminating institution activities, selling courses abroad, and improving the image of the institution. The importance of the image of the institution for the teachers becomes evident. What is more, social media stimulates economical consumption as paper use is reduced and possibilities of networking all over the world reduce travel and save travel time. Cost-effectiveness and the possibility to reach vast numbers of people in almost no time are the features discussed by Kop (2010).

The third super-ordinate theme, concerning the perceived problems of using social media, discloses the problems teachers identify (see Table 3.3):

- Difficulties processing information;
- Difficulties communicating via social media.

The sub-theme of "difficulties processing information" involves such kinds of interference as information overload, time used to find the necessary information and, sometimes, the unreliability of the content. The content unreliability is closely related to what social media theoreticians indicate as an information credibility problem, which could largely be caused by wiki media and its possibilities to copy, correct, use and reuse information. Such processes naturally lead to "information inaccuracies." In addition, social media might be highly distractive, and users might face technical difficulties that limit or postpone user access to the necessary information. The other serious problem is quality of communication, which is connected to difficulties while communicating via social media. Teachers feel somehow reserved and conscious that they have to be "careful on what you make public." Also, teachers claim that lack

Table 3.3 Perceived problems (teacher issues of using social media)

Meaning unit	Abstracted unit	Sub-theme 1	Sub-theme 2	Theme
It takes time to find the correct information	It takes time to find information	Time for finding information	Difficulties processing information	Issues using social media
Unreliability of users generated content	Unreliability of content	Unreliable content		
Sometimes social media is highly distractive	Highly distractive	High distraction		
Too much information, difficult to select	Too much information	Information overload		
Sometimes there are technical difficulties and nervousness	There are technical difficulties	Technical difficulties		
Communication in social networks quite often lacks mutual respect and is full of rude remarks	Communication in social networks quite often lacks mutual respect	Lack of respect in communication	Difficulties communicating via social media	
You have to be careful on what you make public	Careful on what you make public	Careful about publicity		

of respect in communication could be observed on social media. This, however, might be related to what Trowler (2003) identifies that there is a tendency revealing that students may find the teacher authority alien, and students want to learn from persons they like, while teachers still demonstrate behavior of being used to the institutionalized authoritative order.

The super-ordinate theme of change manifestation in teacher case is comprised of two main themes: social-media-generated changes and challenges adapting social media for studies. The theme of social-media-generated changes falls into two sub-themes (see Table 3.4):

• Shifting from text to video information;
• Improved time and resource control.

The shift from text to video is described by McLuhan (2003), who discusses the change of linear textual form into multiple concentric form of infinite crossing of meanings that all interact together; the author points out that multiple forms of media exist in interaction with each other.

The two sub-themes introduce the newly emerging features in university studies prompted by social media use. Hancock (1998) discusses such features, pointing out that they are possibilities of visualizing information; in other words, a visually enriched way of presenting ideas and information. Also, social media creates possibilities to manage time and resources in unprecedented ways unknown before the existence of social media. Teachers even predict the dissolution of physical institutions with the rising possibilities for students to participate in classes from home: "the use of social media will make 'physical organizations redundant' and students could stay at home for classes."

The theme of challenges adapting social media for studies is comprised of two sub-themes: need for incorporating social media into studies and perceived areas for improvement. Teachers recognize the need to embrace social media, making them a part of the university infrastructure and making them a common study tool as well as providing the necessary training for the staff. While discussing the perceived areas for improvement teachers recognize that "the staff are not ready for the change," that people lack necessary skills and ideas of how to use social media in university studies. The necessity for technical and pedagogical support is clearly voiced by the teachers. Ultimately, teachers identify that under the influence of social media, there is an inevitable need to change the institutionalized hierarchy perception. They express the understanding that social media demonstrates the tendency to democratize the teaching and learning process by also dissolving higher teacher hierarchy.

Table 3.4 Change manifestations (lecturers)

Meaning unit	Abstracted unit	Sub-theme 1	Sub-theme 2	Theme
Material will have to move from the text. Social media is still text-based; there would be more video, more multimedia	Material will have to move from the text to more video	Moving form text to video	Shifting to video information	Social media generated changes
E-Systems which save time and resources costs should be created	E-Systems will save time and resources costs	Saved time and resources	Improved time and resource control	
The use of social media will make physical organizations redundant. A lot of the higher education institutions will disappear	The use of social media will make physical organizations redundant	Physical organizations redundant		
One possibility might be to use group video chat to replace the face-to-face meetings we still have. Students could then stay at home for classes	One possibility might be to use group video chat and students could stay at home for classes	Students could stay at home for classes		
The most important challenge is that using social media becomes common in the university. It should not be something individuals do, but has to be the task of every member in the teaching staff	Using social media should become common in the university	Become common in university studies	Need for incorporating social media	Challenges adapting social media for studies
Allow staff time to learn and plan	Time for staff to learn and plan	Time for training staff		
At the moment the staff is not ready for this change	The staff is not ready for the change	No readiness for the change	Perceived areas for improvement	
Not everyone has the necessary skills to take social media into use	Not everyone has the necessary skills			

(continued)

Table 3.4 (continued)

Meaning unit	Abstracted unit	Sub-theme 1	Sub-theme 2	Theme
Teachers are not used to these tools. There is lack of ideas how to use these tools for educational purposes	Lack of ideas how to use the tools for educational purposes			
Too little support technically and pedagogically	Not enough support technically and pedagogically			
Teachers will have to change their idea of being on the higher hierarchy than students	Teachers will have to change their idea of being on the higher hierarchy than students	Teachers have to change hierarchy perceptions		

3.1.1.2 Student Experience

The theme of student use of the social media includes two sub-themes that are similar to those identified for the teacher use of social media (see Table 3.5):

- Contacting people;
- Sharing information.

Table 3.5 Social media application (student use of social media)

Meaning unit	Abstracted unit	Sub-theme 1	Sub-theme 2	Theme
To keep in touch with old friends	To keep in touch with friends	Contacts with friends	Contacting people	Student use of social media
To keep in touch with family members	To keep in touch with family	Contacts with family		
To stay in contact with teachers and connect with their network	To stay in contact with teachers	Contact with teachers		
Meeting new people via social media	Meeting new people	Building new contacts		
Business contacts and opportunities, job searches	Business contacts and opportunities	Business contacts		
Sharing knowledge, sharing experience on study difficulties	Sharing knowledge and experience	Sharing knowledge	Sharing information	
To keep up to date of the studies	To keep up to date	Updating information		

Students remark that they use social media not only for keeping their existing close social contacts such as friends and family—"To keep in touch with friends, family"—but they also use social media for meeting new people. In addition, students recognize that they use social media for study and business contacts: "To stay in contact with teachers, business contacts and opportunities."

The other emerging sub-theme is "sharing information," which embraces "Sharing knowledge, sharing experience on study difficulties or keeping up to date." It demonstrates that students are inclined to not only share knowledge but they also share their experience on how to solve study difficulties, which indicates that collaboration of students takes place on social media.

Students perceive certain advantages of social media (see Table 3.6), which include:

- Increased information exchange;
- Collaboration.

Discussing the increased information exchange, students identify their perceptions that social media enables the users "to reach many people at the same time" and share more information. Concerning the other sub-theme—collaboration—research participants identify such signs of collaboration as working together for study purposes, providing each other with advice and sharing their produced study materials.

The theme of "student-perceived issues while using social media" is comprised of two problematic areas (see Table 3.7):

- Time-consuming;
- Unreliable.

The first area or feature, that social media seems to be time-consuming, includes the voiced perceptions that lack of immediacy determines longer time to get an answer. In addition, students recognize that they spend plenty of time online: "Takes a lot of time online but without it I would not be able to collect all the information

Table 3.6 Perceived advantages (students)

Meaning unit	Abstracted unit	Sub-theme 1	Sub-theme 2	Theme
You can reach many people at the same time; distance doesn't matter	Reach many people at the same time	Better opportunities for sharing information	Increased information exchange	Student-perceived advantages of social media
Social media means sharing more information	Sharing more information			
You can work easily together for the study. We share summaries	Work easily together for the study and share	Better collaborative study	Collaboration	
It's easy to ask other students for their advice	Easy to ask other students for advice			

Table 3.7 Perceived problems (student issues of using social media)

Meaning unit	Abstracted unit	Sub-theme 1	Sub-theme 2	Theme
But it takes longer to get an answer, by telephone you'll get the response immediately	It takes longer to get an answer, not immediately	Longer time for responses	Time-consuming	Student-perceived issues of social media
Takes a lot of time online but without it I would not be able to collect all the information I need	A lot of time online	Increased online time		
Information on social media is not always reliable	Information is not always reliable	Unreliability of information	Unreliable	

I need." The search for the required information and sorting it out, and dealing with the constant information flow, requires time. The other problematic feature is that information is not always reliable, as has been discussed above.

The super-ordinate theme of change manifestations, in the case of students, embraces the theme of student expectations and requirements. It reveals students' concerns and wishes related to the changes in university studies stimulated by the use of social media. The theme of student expectations is comprised of three sub-themes (see Table 3.8):

- Student involvement encouragement;
- Need to address students' issues;
- Need to improve teaching.

Discussing the possibilities for encouraging student involvement, students express their wish that attention would be paid to a more active student participation. They feel it should be increased by encouragement and stimulation as the university itself gains from student active participation: "Students must be rewarded for their contribution; students must participate, university itself gains knowledge from that." Students express their inclination to be actively involved in the study process and their wish to be encouraged. In addition, they would like the institution to take active actions and to address their issues. What is more, real institutional support is expected by students, especially concerning the question of addressing students' problems, which, based on the interviews, include the following areas:

- Addressing psychological issues of students;
- Raising security awareness;
- Provision of familiarization with social media;
- Need to express opinions.

It is obvious that students count on the support of the institution in the process of getting familiar with social media tools for those who experience such a need: "Clearer framework for those who are not familiar with social media." Also, students

Table 3.8 Change manifestations (students)

Meaning unit	Abstracted unit	Sub-theme 1	Sub-theme 2	Theme
If you want your students to be active, they must get something in return. Students must be rewarded for their contribution	Students must be rewarded for their contribution if you want your students to be active	Student activity increased by rewards	Student involvement encouragement	Students expectations and requirements
Students must participate; university itself gains knowledge from that	Students must participate	Necessity of student participation		
More emphasis should be put on the social psychology of the student who "lives" in the social media and what is their relationship with society	More emphasis on the social psychology of the student who "lives" in the social media	Addressing psychological issues of students	Need to address students issues	
More security aware social media, more projects leaded by psychology aware people	More security aware social media	Raising security awareness		
More information about security in social media	More information about security			
Clearer framework for those who are not familiar with social media	Clearer framework to familiarize with social media	Provision of familiarization with social media		
I would like everyone to be braver and express their real opinions without fear, I'd also like those whose opinions are addressed to, to respond to them	I would like everyone to express their real opinions without fear	Need to express opinions		
There should be online consultations, question sessions with the teacher	Online consultations, question sessions with the teacher	Promoted online teaching	Need to improve teaching	
Seminars should be put on the platform to give the way to those absent to have the material of the lessons at home	Seminars should be put on the platform to have the material at home			
Teachers are often aware what social media's world concern, actually sometimes they are even more acknowledged than us, for what e-learning concerns for instance, but they remain on traditional teaching using old tools	Teachers are often aware and acknowledged but they remain on traditional teaching using old tools	Persistence of traditional teaching		

expect that the institution would provide psychological aid and security while using social media: "More emphasis should be put on the social psychology of the student who 'lives' in the social media and what is their relationship with society; more security aware social media." The interviews with the students disclose the concern about psychological disturbances that students experience while using social media.

Another concern identified by the student research participants is the expression of opinions without fear, which seems to be related to identity issues. Bauman (2011) states that the interplay of multiple identities becomes an accepted norm in social media; opinions become just a matter of choice that is determined by ever-changing whimsical fashion. Opinions are compared to clothes, which being fashionable for one season get completely out-of-date the next season. The consumer world inflicts a stigma of being old-fashioned, so people would experience fear taking risks of expressing an unfashionable opinion.

Moreover, students reveal their wish for improvement of teaching processes. They acknowledge that, for the most part, teachers are well-aware of social media and that sometimes teachers know about social media even more than the students themselves, but the use of traditional teaching persists: "Teachers are often aware and acknowledged but they remain on traditional teaching using old tools." Students voice the need for online teaching promotion, and they point out that they would like more activities online such as consultations, question sessions and all the materials available online: "There should be online consultations, question sessions with the teacher, lessons should be put on the platform to give the way to those absent to have the material of the lessons at home." The student voice reveals that it might be the case that traditional teaching has the tendency to persist in mixed-mode teaching rather than in e-learning and distance learning where the entire courses and teacher consultations are available online.

3.1.1.3   Administrator Experience

In the super-ordinate theme of social media application, the theme of the administrator use of social media comprises three sub-themes (see Table 3.9):

- Information exchange;
- Communication;
- Promoting student/client engagement.

The sub-themes indicate the main purposes for which administrators use social media. They use social media for information exchange, which includes exchanging ideas and sharing comments; additionally, they use social media for communication and promoting student/client involvement. The sub-theme of student/client involvement indicates that administrators are interested in the institution clientele and that it is directly connected with administrator job matters. Administrators use social media "to acquire new students... for building a society of continuous education customers and prospects."

Table 3.9 Social media application (administrator use of social media)

Meaning unit	Abstracted unit	Sub-theme 1	Sub-theme 2	Theme
Use social media for exchanging opinions, expressing my ideas	Use it for exchanging ideas	Exchanging ideas	Information exchange	Administrator use of social media
I write some comments to the articles and answer the other people's comments	Write some comments to the articles	Sharing comments		
Use social media for communicating with institute members	Communicating with institute members	Communicating with colleagues	Communication	
Use it for building a society of continuous education customers and prospects	Building a society of continuous education customers	Forming customer group building clientele	Promoting student/client engagement	
It's a good way to acquire new students	A good way to acquire new students	Acquiring new students		
Easier to reach future students	To reach future students	Reaching new students		

The theme of administrator-perceived advantages of social media (see Table 3.10) involves:

- Institution promotion;
- Information exchange promotion.

The sub-theme of institution promotion is closely related to more opportunities of advertising and institution image improvement; for example, administrators indicate that "social media gives more opportunities for advertising ourselves, increases visibility of institution." Information exchange and promotion comprises such points as speed of sharing information, increased communication, independence of time and place, and activation of all sensory channels.

The administrator-perceived problems (see Table 3.11) include:

- Time consumption;
- Training necessity;
- Limited learning control.

Administrators indicate that social media takes more time in certain aspects; for example, "more time for managing the relationships and the communication." It also takes time for mastering it: "It takes time to take social media into use." What is more, administrators see the necessity for staff training by recognizing that "not everybody is ready to use social media; there is lack of technical skills." An additional problem administrators identify is limited learning control. Administrators demonstrate

Table 3.10 Perceived advantages (administrators)

Meaning unit	Abstracted unit	Sub-theme 1	Sub-theme 2	Theme
Social media gives more opportunities for advertising ourselves, to show the positive aspects of organizing the study process	Gives more opportunities for advertising ourselves	Advertising opportunities	Institution promotion	Administrator-perceived advantages of social media
It increases visibility of institution	Increases visibility of institution	Improving institution image		
It means reputation, keeping pace, student flexibility and engagement, new markets, support overseas students, partner institution, widening access	Reputation and widening access			
It ensures speedy information sharing	Speedy information sharing	Increased information sharing	Information exchange promotion	
It increases communication and sharing	Increases communication and sharing			
Social media provides independence of time and place	Independence of time and place			
It means more visuality, all senses into play	All senses into play	Activation of sensory channels		

their concern about the way social media influences the teaching/learning process: "It makes it complicated, control of learning results, systems to control the students; little institution control is possible." Foucault (1998) indicates that educational institutions show ritualized practices regarding discipline and control. Similar inclination to control could be heard in administrator comments voicing their disposition toward a controlling approach. According to the insights of Deleuze (1987), however, the hierarchical democracy loses its prevailing positions in social media culture, and instead democracy without any visible center manifests itself in the mosaic social media world. Administrator comments disclose the ongoing change discussed by Deleuze (1987); the symptoms of the change are difficult to control but the interviews with administrator research participants reveal their wish to control. On the other hand, Personal Learning Environment (PLE) could be an innovative solution which can stimulate autonomy and individual organization and gives preference to decentralized learning strategies independent of educational institutions. However,

Table 3.11 Perceived problems (administrator issues of using social media)

Meaning unit	Abstracted unit	Sub-theme 1	Sub-theme 2	Theme
Social media "eats" time, takes time away from quiet reflection	Takes time away from quiet reflection	Time for managing social media activities	Time consumption	Administrator-perceived issues
More time is needed for managing the relationships and the communication social media induce	More time for managing the relationships and the communication			
It takes time to take social media into use	Takes time to take into use	Time for mastering social media		
There is lack of technical skills	Lack of technical skills	Need for skills	Training necessity	
Not everybody is ready to use social media	Not ready to use social media			
Teachers need more guidance considering the use of social media	Teachers need more guidance	Need to guide teachers		
Many challenges: learning quality control, level of education	Challenges for learning quality control	Complicated learning quality control	Limited learning control	
Social media will change the way we learn, probably make it more natural but less controllable, defining learning goals could be more difficult, examination of knowledge could be difficult	The way we learn becomes less controllable			
It makes it complicated, control of learning results, systems to control the students	Complicated control of learning results, systems to control the students			
Little institution control is possible	Little institution control			

it requires content production which is a time-consuming activity for both students and teachers (Rodrigues and Lobato Miranda 2013).

The theme of institution support falls into the super-theme of change manifestation. The main sub-themes revealed in the course of the interviews (see Table 3.12) are:

Table 3.12 Change manifestations (administrators)

Meaning unit	Abstracted unit	Sub-theme 1	Sub-theme 2	Theme
The institution provides an infrastructure on which the teaching staff can make available/choose social media-based services to be used for teaching/learning	The institution provides an infrastructure for the teaching staff	Institution provided infrastructure	Infrastructure development	Institution support
University provides financial help for social media, technical and infrastructure support	University provides financial help	Provided financial help		
University is offering training sessions to staff	University is offering training sessions	Provided training	Staff development	
University provides guidance for teaching staff	Guidance for teaching staff			
University provides advice and assistance for students	Advice and assistance for students	Provided student assistance	Student assistance	

- Infrastructure development;
- Staff development;
- Student assistance.

Infrastructure development shows that institutions support the use of social media by providing financial help as well as technical and overall infrastructure improvement. It demonstrates that universities recognize the importance of social media and express their recognition by providing material support. In addition, universities provide training and guidance for the staff who are the main actors promoting social media use in the institutions. What is more, universities take into account their students and provide guidance and assistance for them.

3.1.2 Key Emerging Contradictions

The established super-ordinate themes allow for the analysis of the contradictions faced by users of social media in different universities participating in the ISTUS project across Europe. Our research highlights the contradictions of social media use in university studies faced by the three different groups of university studies participants interviewed in the framework of the project research.

The first clearly expressed contradiction that could be elicited from the super-ordinate theme of the perceived problems of using social media is related to infor-mation literacy. The extensive use of social media for sharing information and the difficulties processing information, the occasional unreliability of information and the abundant time spent on social media create a contradiction, a solution to which could be the development of the necessary skills of information literacy. Teachers discuss difficulties processing information, which include impediments like infor-mation overload, time used to find the necessary information and, sometimes, the unreliability of the content. Similar difficulties are disclosed by students, who reveal that the constant information flow, searching for the necessary information and sort-ing it out, requires a certain amount of time. In addition, the information found on social media is not always reliable. Administrators also state that social media is time-consuming: "More time for managing the relationships and the communica-tion." In addition, mastering social media as well takes time: "It takes time to take social media into use." The three groups of the research participants, teachers, stu-dents and administrators, disclose the problems of processing information collected on social media.

In this way, information literacy emerges as a key skill necessary for successful information searching and processing on social media. The definition of "information literacy" provided by the American Library Association (1989) identifies that infor-mation literacy is a set of abilities enabling individuals to recognize when information is needed and have the ability to locate, evaluate and use effectively the necessary information. Information literacy stands out as increasingly important because of the fast technological change and the abundant availability of information resources. People are impacted by diverse information modes and sources and have to manage information and use it effectively due to the needs of their study, work, or personal lives. Universities as educational institutions face the need to incorporate information literacy across curricula; such a need becomes essential and requires certain efforts from the faculty and administrators.

Another contradiction is related to the inherent democratic nature of social media. As Deleuze (1987) observes that the prevailing positions of the hierarchical democ-racy seem to have been lost in social media culture instead democracy without any visible center has manifested itself in mosaic social media world. Teachers feel lack of respect in social media communication environments. Administrators express their concern about the control of the learning process, whereas students advocate for more student participation—"Students must participate, university itself gains knowledge from that"—and they would like to be considered as active participants in the process of knowledge creation. Students, being profound social media users, are used to the mosaic world of hypertext and express their confidence that they are able to contribute to knowledge construction at university. Contrarily, teachers and administrators disclose their predisposition to get student respect and control the university study process. This seems to be related to what Foucault (1998) voiced—that educational institutions have ritualized their practice regarding discipline and control; however, social media use in university studies acts as a driving force toward democratization of educational processes.

Identity issue, which students point out, reveals a contradiction connected to the possibility to express opinions freely. It resonates with Bauman's (2011) idea that opinions are like clothes which have the feature of being fashionable one season and getting completely out-of-date next season. What is more, being old-fashioned is considered to be a stigma in the consumer world, so students feel cautious that they might express an unfashionable opinion and might be driven to refrain from expressing an opinion at all. Personal identity becomes an object to be hidden or not expressed explicitly, or lost in the mosaic game of media. Constant identity change, dismissal of the old ones, and seeking new expressions become a must, which stimulates and forces the play of identities.

The identity issue is closely connected to security problems on social media because identity processing becomes important as people voice their concerns about safe identity processing and means of prevention of identity theft. The privacy problem adds extra anxiety concerning social media use since many users tend to be not careful about what they express and expose in their social media space. Research participants voice the problem: "You have to be careful on what you make public."

The change of university study processes provokes further contradictions. Students voice their preferences for changing in teaching: "Teachers are often aware of what social media's world concerns; actually, sometimes they are even more knowledgeable than us—for what e-learning concerns, for instance—but they maintain traditional teaching using old tools." According to Prensky's (2014) "VUCA" (variability, uncertainty, complexity, and ambiguity) term, which stresses the growing complexity of our learning and living environments, contemporary teachers and students face variability of the education technologies and complexity of the educational and developmental environments. University teachers are faced with the necessity of embracing ever-changing and growing technologies and applying them in teaching.

Similarly, the problem of accepting the digital reality of social media is discussed by McLuhan (2003), who admits that the era of mechanistic and linear philosophy prevalence has come to an end; linearity has been replaced by the digital era with its simultaneity and concentric nature with infinite crossing of planes where all types of media constantly interact with each other. Nevertheless, it takes effort and time to master the technologies and adapt to changes. The need for training to embrace social media within teaching and learning is voiced by all the three groups of research participants: teachers, students, and administrators. What is more, students express a certain need to gain guidance for using social media in their learning: "Clearer framework for those who are not familiar with social media," acknowledging that there are certain differences in social media mastering skills among students.

Concerning learning, social media induces a closer fusion between formal and informal learning. On one hand, students receive more possibilities for working on their own or with other students without teacher guidance in activities like sharing knowledge or collaboration: "Sharing knowledge, sharing experience on study difficulties." On the other hand, students voice their expectation to be guided and evaluated by teachers: "There should be online consultations, question sessions with the teacher." According to the connectivism theory introduced by Siemens (2004), the importance of informal learning increases as it becomes a significant part of learning experience. Technology changes the ways in which people process and manage information, and learning could be viewed as the process of connecting specialized information sets, prioritizing and choosing what to learn, and being able to identify the connections among multiple fields, ideas, and notions.

In such a process of dealing with multiple choices and connecting multiple ideas, teachers may become facilitators or advisors as to which information and ideas to select by showing their students the basic grand theories and the ideas supporting them. Teachers may assist their students by helping them glide more easily in the multifaceted ocean of information; yet, students voice their preference more for guidance rather than control such that the role of institutional culture and the role of teachers should move toward the more guiding paradigm rather than the controlling one. Social media inevitably envisions a more democratic and less formalized way of teaching and learning; however, institutionalized hierarchy is still a hurdle for administrators to overcome as they voice their caution about the less controlled learning process.

Another controversial area discussed by the teachers is the redefinition of educational institutions. The restructuring and fragmentation of the postmodern world forces the modern human to make independent decisions and construct a personal reality and to reconstruct the personal, social, and working world (Glastra et al. 2004). Teachers voice their supposition that educational institutions as they are might undergo certain changes: "The use of social media will make physical organizations redundant. A lot of the higher education institutions will disappear" or "One possibility might be to use group video chat to replace the face-to-face meetings we still have. Students could then stay at home for classes." As McLuhan (2003) points out, the age of digital technologies has redefined the nature of work, more people were freed from repetitive mechanistic work, and in this way people gained more possibilities to participate in society creatively; work has become decentralized with a multitude of opportunities. Although social media provides more opportunities for advertising institutions and attracting new clients—as the administrators point out, this "gives more opportunities for advertising ourselves, to show the positive aspects of organizing the study process" and "reputation, keeping pace, student flexibility and engagement, new markets, support overseas students, partner institution, widening access"—at the same time, social media modifies the nature of teacher work and the nature of education itself.

Administrators recognize that, "It takes time to take social media into use" and "Teachers need more guidance considering the use of social media", and administrators state that they are ready to provide the training and the guidance: "University is offering training sessions to staff and provides guidance for teaching staff and advice and assistance for students." The process of embracing social media is continuing, and hopefully it will set the new ways of expression of human creativity in society.

3.1.3 Emerging Positive Effects

As established by the research, all the three groups of research participants—teachers, students, and administrators—use social media for communication and sharing information. This is in line with the observation of Kietzmann and Hermkens (2011) about functional blocks of social media wherein they state that "communication block" represents the ways in which social media users converse with each other for various reasons. Research participants observe that they communicate with their friends, colleagues, and organizations for personal and professional reasons. Teachers not only communicate with their personal contacts, they also communicate with their students and other organizations for educational purposes. Students use social media not only for educational purposes, they also search for business contacts. Naturally, administrators are interested in representing their institution and attracting new clients. Another block—the information sharing block—also seems to be very important for the research participants. They share various ideas and information on organizations and studies: Teachers share various visual and other study materials with their students, students collaborate on their studies and exchange information on study difficulties and try to solve them, and administrators share ideas and also participate in the process of "building a society of continuous education customers and prospects"—informing their prospective customers, presenting study opportunities, and spreading information on the institution.

The super-ordinate theme of perceived advantages identifies positive social media effects concerning technology incorporation into university studies. According to Kop (2010), modern information technologies such as social media may positively influence education. Kop (2010) identifies certain features appearing in education due to the use of social media, and the research participants experience the same features:

- Increased communication;
- Sharing information;
- Possibility to reach vast numbers of people almost instantly;
- Information visualization;
- Wide access to information;
- Simulation (various types of projects that enable the use of various sensory channels);
- New forms of creativity;
- Economical nature, saving resources.

Teachers recognize that wide access to information while using social media increases the possibilities to share information, to exchange ideas and to get updated. Teachers voice it as: "Getting up to date about what other people think." Another feature identified by the research participants is a chance to reach broad audiences: "Reaching more students and communicating in a better way." Teachers recognize that many more students could be reached and, what is more, some features improve communication quality; for example, the possibilities of instant feedback and discussions. Teachers point out the importance of promoting the institution, disseminating educational activities, and representing the institution better by creating and promoting a better image of an institution and making it attractive to students. Additionally, teachers speak about the economical feature, saying that social media helps to save paper and travel time and, at the same time, makes education more accessible by making it more economical: "Networking all over the world, less travel time."

Students also recognize wide access to information. Multiple sources of information could be used and multiple users could be engaged, which improves working together and collaboration in knowledge exchange and creation. New knowledge creation in a collaborative way is connected to new forms of creativity induced by the process of creativity becoming not only an individual secluded pursuit but also manifesting itself in the process of sharing, working together, collaborating, and creating in multiple groups of social media users. Administrators also recognize the importance of wide access to information. They identify such features as increased communication and sharing and accessing information independently of time and place.

All the features established by the research are important, and they all influence university studies. Participants of educational processes may rely on enhanced communication via social media, enhanced access to information resources, engagement of all the sensory channels, and collaborative creation of new digital content. What is more, education participants positively view the mentioned features and recognize them as moving university studies toward improvement, toward new quality and new envisioned dimensions.

3.2 Inductive Qualitative Study with a Phenomenological Approach at Home Institution

The findings of the inductive qualitative research with a phenomenological approach include the descriptions of the emerging themes and their clustering into super-ordinate themes. The analysis allows us to establish three intertwining layers. The first layer includes major contradictions and positive effects of social media use in university studies induced by the overview of themes and super-ordinate themes. The subsequent layer is derived by applying a phenomenological approach and consists of the deciphered existentials of lived experience indicated by van Manen (2014): relationality, spatiality, and temporality. Finally, at the deepest level of the analysis the two key super-ordinate themes are highlighted: lecturer creativity expression while using social media in university study processes and the pedagogical relationship between lecturers and students while using social media.

3.2.1 Findings of the Inductive Qualitative Study with a Phenomenological Approach at Home Institutions

The analysis of the interviews was used to identify the meaning units in the research participant interview material. After that, the meaning units were condensed into sub-themes, which were then linked into clusters to form the main themes and super-ordinate themes. All the procedural steps for defining super-ordinate themes are presented in the tables below. The tables are organized according to the three groups of research participants: teachers, students, and administrators, reflecting the experiences of social media application in university studies for each research participant group accordingly. Four common super-ordinate themes for all the three groups of participants have been identified: social media application, perceived advantages, perceived problems, and change manifestations. Two additional super-ordinate themes for the teacher group also emerged: pedagogical relationship and teacher creativity, and one additional super-ordinate theme for the administrator group: reservations.

3.2.1.1 Teacher Experience

The first super-ordinate theme includes the themes of social media use by teachers, students, and administrators. Table 3.13 contains the information on the main ways in which teachers use social media.

The sub-themes reveal that teachers identify the following purposes of social media use:

- As a source of information;
- For sharing information;
- For communicative purposes;
- For interactive activities;
- For stimulating student creativity;
- For advertising purposes.

Teachers share information with their students and colleagues; sometimes they even give advice and teach their colleagues to use social media: "And I propagate social media and give advice for everyone and teach all my colleagues." They also use social media as a source of information, searching there for materials for their lectures, as well as for personal interests and use. As social media is designed for socializing, naturally teachers use social media for communication both with work-based contacts—students and colleagues—and personal ones. In their work, teachers use interactive tasks that they create themselves or find ready-made in social media: "Mainly I create and use interactive tests, I use interactive vocabulary tasks and other tasks" and "There is a world platform for learning Spanish, there are plenty of inter-active tasks, I use it a lot." Teachers are also aware of stimulating student creativity

Table 3.13 Social media application (teacher use of social media)

Meaning unit	Abstracted unit	Sub-theme 1	Sub-theme 2	Theme
I compile everything and present to the students and for them it's a safe understanding, lecture plan, and material	Present to the students lecture plan and material	Sharing material with the students	Sharing information	Lecturer use of social media
Use it as a library, attach files for students	Attach files for students			
Use for sending material for watching videos	Use for sending material			
I try to do everything in the Internet, fill in and prepare documents, various reports	Prepare documents	Sharing documents		
And I propagate social media and give advice for everyone and teach all my colleagues	Propagate social media and teach colleagues	Teaching colleagues		
Use media for searching information	Use for searching information	Information search	Information source	
Use social media to find out about the events	Find out about the events			
On the Internet we watch what interests us, read articles, watch videos	Watch what interests us			
I often check the new photographs of my contacts	Check the new photographs of my contacts	Contact information update		
I blocked my Facebook account but now I want to unblock it because I would like to see what my son is doing on Facebook	I would like to see what my son is doing	Family member information		
I use social media for socialization, can communicate with the students in a distant way	Can communicate with the students in a distant way	Communication with the students	Communication purposes	
I use Facebook for communication with friends	Use for communication with friends	Communication with friends		
Use for communication, to connect wherever I am and whenever	Use for communication, to connect			

(continued)

Table 3.13 (continued)

Meaning unit	Abstracted unit	Sub-theme 1	Sub-theme 2	Theme
Mainly I create and use interactive tests, I use interactive vocabulary tasks and other tasks	Create and use interactive tasks	Use interactive tasks	Interactive use	
There is a world platform for learning Spanish, there are plenty of interactive tasks, I use it a lot	Use interactive tasks from platforms			
Use various forums and interactive tasks	Use forums and interactive tasks			
We create blogs with the students for particular projects, particular tasks students create blogs, and here appears creativity aspect	Create blogs with students for particular projects	Blog creation with students	Stimulate student creativity	
I also administer the Facebook page of university academy of philology so I post here interesting information to attract students, to attract new participants	Post interesting information to attract students, new participants	Attracting new participants	Advertising purposes	

while using social media, so they purposefully use social media for certain activities to develop student creativity: "We create blogs with the students for particular projects, also for particular tasks students create blogs, and here appears creativity aspect." Some teachers are responsible for organizing broader events in the faculty or the whole university, and then they use social media for advertising purposes: "I also administer the Facebook page of the university academy of philology, so I post here interesting information to attract students, to attract new participants (Table 3.14)."

In the super-ordinate theme of perceived advantages of social media use, the theme of teacher-perceived advantages includes numerous sub-themes:

- Promotion of information exchange;
- Broader view of a person;
- Media attractiveness;
- Availability across time;
- Extended space limitations;
- Increased student involvement;
- Economical;
- Institution promotion.

The sub-theme "promotion of information exchange" includes accessibility of teaching materials and information exchange as well as quick access to "everything" and easy student work monitoring and assessment. As the research participants say:

Table 3.14 Perceived advantages (teachers)

Meaning unit	Abstracted unit	Sub-theme 1	Sub-theme 2	Theme
For me it's a space where I can get information easily, to ask and get answers easily	I can get information easily, to ask and get answers easily	Accessibility of information exchange	Promotion of information exchange	Perceived advantages of social media
At present it is the safest for me, I mean that can always reach my teaching material and it doesn't disappear anywhere	I can always reach my teaching material	Accessibility of teaching material		
And you can save everything, nothing what you do disappears, everything is archived and you can save big quantities of your activities	Can save big quantities of teaching material			
Speed, a quick access to everything, quick search, quick to reach people	a quick access to everything, to search and people	Quick access		
I know how quickly renew it, I have mastered all the functions in blogs	Can quickly renew the material			
Independent work is a part of the course and students get credits for it, so it has to be assessed and this is a perfect way of assessment of student independent work because we can give not only test, there is a wiki in Moodle and we can give common project for students to prepare	A good way to monitor and assess student work	Student work monitoring and assessment		
I have less to check, less to monitor students in Moodle environment than if they did the tasks in the paper variant, it saves my time and my attempts	Less checking of student work			
From the photographs, from everything you can understand, you can find out the opinions of other people, you can find out more than communicating in an ordinary way, because the person may not openly tell you what he likes, what music, what creative arts…	Can find out the opinions of the people you communicate with and can find out more than in a simple communication	Can find out more about personality	Broader view of a person	
In Facebook you see a broader view of a person, maybe a person reveals in a different way	Can see a broader view of a person			

(continued)

Table 3.14 (continued)

Meaning unit	Abstracted unit	Sub-theme 1	Sub-theme 2	Theme
First of all I paid attention to what language people use in their writing, then what moods they express	You can see the language of people, their moods			
You can find out what is going on, what is important for people at the moment	Can find out what is important for people at the moment			
What is more I have a wall wisher which is similar to wiki but more playful	It is more playful	Media playfulness	Media attractiveness	
Social media can provide the information and visual and audio, so they search for various forms of information	Information and visual and audio	All senses involved		
There is Moodle environment prepared and I am a non-editing moderator there and everything is easy and simple for me	Easy and simple to use social media	Easiness of social media use		
It is so easy to master it, later you get used to it and start managing how you get the information, what you want to present and how you want to present the information	Easy to master, get used to and master how to get the information			
There are possibilities to communicate, relax, stress relief, focusing on other things more productively	Relaxation and stress relief	Relief possibilities		
In a way kind of privacy appears, in fact you get into public space but as everyone uses it, you remain unnoticed	Privacy of being in public spaces as media involves everyone	Privacy while being in public		
There are no time or place limits, students can connect whenever, teachers can connect whenever	No time limits for connecting	Extended time limits	Availability in time	
Very convenient when you get ill you can connect to your students from a distance	You can connect to your students from a distance	Connecting from a distance	Extended space limitations	
Some time ago I had to go to work to meet the heads of departments to solve some issues, now when there is email, you don't need to go in order to reply to the questions or send a document	Don't need to go to work to solve some issues	No need to physically cross space		

Table 3.14 (continued)

Meaning unit	Abstracted unit	Sub-theme 1	Sub-theme 2	Theme
The search for information becomes quicker, earlier when you needed some information first you checked resources at home, then you would go to the library, and not always could find the information, now the Internet allows you to find the necessary information very quickly	To search for information you don't need to go to the library			
Some students are shy and they perform and reveal themselves better while working on their own in social media environment	Shy students perform better in social media environment	Encouragement of student openness	Increased student involvement	
My students admitted that for them being behind the technology screen allows them to open up, to communicate more sincerely	Students admit that technology allows them to open up			
What is more, they get the answer immediately, if it's correct or wrong, where there is a mistake and they can correct themselves immediately and they get involved in trying to get more points	Students get feedback immediately and they get involved	Accessibility of feedback		
For me it's better than to carry lot of papers around	No need for a lot of papers	Saving resources	Economical	
It's clear that you save on travel, you can participate in conferences not traveling abroad	Save on travel			
Information about our university is disseminated in such a way and we can attract new students	Information about our university is disseminated	Information dissemination	Institution promotion	

"For me it's a space where I can get information easily, to ask and get answers easily" and "Speed, a quick access to everything, quick search, quick to reach people." Even the research participants' language is intended to convey the experience of speed by using short chunks and repetition. As Gadamer (1999) wrote in his texts about human and language, a human can express him/herself and his/her experience through language. The research participants also report that it is easy to monitor student activities with social media: "This is a perfect way of assessment of student independent work because we can give not only test, there is a wiki in Moodle and we can give common project for students to prepare" and "I have less to check, less to monitor students in Moodle environment than if they did the tasks in the paper variant, it saves my time and my attempts."

The next sub-theme, "broader view of a person," reveals that in social media we can find out people's opinions, hobbies, and interests;—in other words, information that we maybe would not ask or would not otherwise know, information they share that reveals the personalities: "From the photos, from everything you can understand, you can find out the opinions of other people, you can find out more than communicating in an ordinary way, because the person may not openly tell you what he likes, what music, what creative arts." As one of the research participants says, "First of all I paid attention to what language people use in their writing, then what moods they express." Again, citing Gadamer's (1999) insight, the language reveals the human being.

The sub-theme "media attractiveness" includes a number of reasons: media playfulness, involvement of all senses, easy mastering of social media and easy use of social media, relief possibilities and privacy while being in public—in other words, being in public but behind the computer screen. I myself, while working on my thesis, when I need a break I visit Facebook to relax myself, first to get visual information instead of written, to change the topic and—yes, in fact, it is like going out without leaving home—sometimes you can even meet a friend and chat for a while. One research participant reveals her lived experience as: "Indeed it seems that spending time in Facebook writing comments is like a crime but if you yourself approach social media and start actively using it, you start understanding that it becomes like addiction, like smoke, when you get indignant when others smoke, but if you start smoking yourself, it is already not only smoke, it is a possibility for communication, stress relief, shifting your attention."

The following two sub-themes—availability in time and extended space limitations—could be viewed together as they deal with time and space. Van Manen (1997) distinguishes four existential themes characteristic to phenomenology: temporality, spatiality, corporality, and relationality, so no wonder that time and space appear in the sub-themes. What is more, Moore (1993) indicates that online interaction is transactional in distance and time. Mason and Rennie (2008) state that the type of technological medium can support student possibility of access through time limitations and various locations, in fact providing access to many users. The research participants experience the accessibility across time and space as almost unlimited possibility: "There are no time or place limits, students can connect whenever, teachers can connect whenever" and "Some time ago I had to go to work to meet the heads of departments to solve some issues; now when there is email, you don't need to go in order to reply to the questions or send a document."

Another sub-theme—increased student involvement—reveals itself through student openness, encouragement, and availability of feedback. The research participants observe that sometimes student potential is revealed better through the medium of social media: "Some students are shy and they perform and reveal themselves better while working on their own in social media environment." Chickering and Ehrmann (1996) indicate that technologies can strengthen faculty interactions with all students but especially with shy students who are reluctant to ask questions or challenge the teacher directly. As well, Ellison, Steinfield, and Lampe (2007) speak about social capital acquired by students using social media. Availability of immediate feedback is the other sub-theme related to student involvement. As the research participants indicate, the immediacy of feedback stimulates students' learning and involvement: "What is more, they get the answer immediately, if it's correct or wrong, where there is a mistake and they can correct themselves immediately and they get involved in trying to get more points."

The sub-theme of institution promotion shows that teachers are aware that dissemination of information about the university helps to attract more students, which is important for teachers themselves. The sub-theme of economical use reveals that paper resource use becomes unnecessary and travel expenses are saved as well. Similarly, Hancock (1998) alludes to the cost-effectiveness of social media and its possibilities for reaching vast audiences in almost no time (Table 3.15).

The third super-ordinate theme reveals the issues perceived by the teachers while using social media:

- Difficulties processing information;
- Addiction;
- Difficulties with technological supply;
- Time-consuming processes;
- Fragmented communication;
- Cautiousness about privacy.

The sub-theme of difficulties processing information includes such impediments as scattered information, advertisement intrusion, material loss, distraction by unused functions, unreliable information, and the need to deal with a non-stop information flow. In fact, many researchers speak about fragmented information, the problems with information reliability, which imply that social media also express many false truths or lies, much information that is not always true, constant information flow and the necessity of dealing with the information flow, the ability to successfully navigate the information and choose what is appropriate for the individual user in the given circumstances. Rheingold (2010) identifies five types of information literacies with which social media users should equip themselves in order to become successful social media users; since the literacies are not inherent they must be acquired to empower social media users.

Another sub-theme—addiction—is connected to what Bauman (2011) speaks about: that we in our media-permeated world get caught up without noticing how media starts ruling our lives. As one study participant says, "If you master social media and you start using it actively and you start understanding that it is becoming like an addiction." Information technologies are not an addiction like a bad habit, but it is media-permeated life we live and already we cannot imagine our lives without social media. My personal experience as well reveals having got used to social media as now I feel uncomfortable if the Internet disappears—sometimes it even causes stress. Here I remember the movie *The Matrix*, which reveals that human existence is so intertwined with technologies that sometimes it is difficult to say if technologies govern us or if we use technologies as tools.

Table 3.15 Perceived problems (teacher issues of using social media)

Meaning unit	Abstracted unit	Sub-theme 1	Sub-theme 2	Theme
First it would irritate me with kind of aggressive and scattered information dissemination	Would irritate me with aggressive and scattered information dissemination	Scattered information	Difficulties processing information	Issues using social media
While experimenting I lose the material and cannot find it, so I dislike this	Lose the material and dislike the loss	Loss of material		
And I dislike all the visual advertisement intrusion	I dislike advertisement intrusion	Advertisement intrusion		
Moodle environment I use in a reserved way because it interrupts me, there are so many functions which I don't use and they distract me a lot	Interrupted and distracted by many functions which I don't use	Distraction by unused functions		
The information flow is non-stop and you have to live with it	Living with non-stop information flow	Dealing with non-stop information flow		
The abundance of information, how to find, how to deal with it	Dealing with the abundance of information			
They do not look critically at their products, it's a fact, they upload everything with all their mistakes	Upload of information with mistakes	Unreliable information		
You can get lost in the abundance of the information, sometimes you cannot trust the sources of information	Sometimes you cannot trust the sources of information			
If you master social media and you start using it actively and you start understanding that it is becoming like an addiction	Social media is becoming an addiction	Becoming addicted	Addiction	
Freedom is this that you need to have a computer and the Internet connection	Necessary a computer and the Internet	Necessity of computers and Internet	Difficulties with technological supply	
If I have an online lecture in the morning, so the students should have their own computers or they have to have access to computers at university library, so there is a question if there are enough computers for students to access	For online lecture students need to have access to computers	Students need access to computers		

(continued)

Table 3.15 (continued)

Meaning unit	Abstracted unit	Sub-theme 1	Sub-theme 2	Theme
It's difficult to administer tests because not all the computers work, for example, there are nine computers here and two of them don't work and there are 16 students, so I often give them a paper test	Difficult to administer tests as not all the computers work	Necessity for computers		
It often happens to me that plan to watch movie excerpts for GIST and out of a sudden the projector switches off and then you have to improvise to think of other activities to be done				
Certainly at the beginning until you get used how to apply social media, it seems difficult and takes time but you have to invest your time when you are learning	You have to invest your time when you are learning	Understanding about the time input	Time-consuming processes	
I spend plenty of time in order to prepare the material	Plenty of time to prepare material			
Certainly, it takes time for me to upload the test	Takes time to upload the test			
At the initial stage there is a consumption of lecturer time and efforts	Time consumption at the initial stage			
Sometimes you spend hours in front of computer screen looking for information and after that you cannot understand what you were doing at the computer, as if it eats up your time	Spend time looking for information			
You save on fuel to go to work but don't save time, I had to upload additional tasks, and when I had a lecture where there were 30 people, it was a great stress for me, because I got so many questions and while I am answering one question I see that a list of questions is awaiting and I see that a student is already waiting for 10 min, I was all sweaty	You don't save time, have to upload additional tasks, takes time to answer students questions			
Direct contact with the students is disappearing in social media	Direct contact with the students is disappearing	Less direct contact	Fragmented communication	

(continued)

Table 3.15 (continued)

Meaning unit	Abstracted unit	Sub-theme 1	Sub-theme 2	Theme
Live communication is becoming scarce	Live communication is scarce			
It irritates me when you are talking to a person and he/she is connected to his iPad or something and he/she is constantly on Facebook, it seems he/she is not communicating with you	If a person is connected to a network it seems he is not communicating live	Disrupted live communication		
Personally I do not use Facebook because I do not know how my personal data are used	I don't use because I don't know how my personal data is used	Insecurity about personal data use	Cautiousness about privacy	
Somehow I became afraid of all the networks, because Google search can see what you are doing, all my moves in the digital space are registered	I became afraid of all the networks because all my moves are registered	Fear of being followed		
Maybe it's my character, I don't like showing my photographs, telling about myself, it's too personal	I don't like showing my personal information	Reluctance to reveal personal information		

The next sub-theme reveals that there still exists some tension in the sphere of technological supply. As the research participant says, "It's difficult to administer tests because not all the computers work, for example there are 9 computers here and two of them don't work and there are 16 students, so I often give them a paper test." Teachers then have to choose other methods that are applicable without employing technologies.

Another sub-theme is connected to time—one of four existential themes identified by van Manen (2014)—and here we see that technologies are perceived as time-consuming by the research participants. In the super-ordinate theme of advantages, technologies are perceived as available across time and social media is spoken of as if it helps to cross time limitations, while here in the super-ordinate theme of issues social media is perceived as requiring time. Time is therefore experienced in a manifold fashion as being available and as being consumed in big quantities. The research participant states: "Certainly at the beginning until you get used how to apply social media, it seems difficult and takes time but you have to invest your time when you are learning." It takes time to master social media, time "to upload tests," time "to prepare material," and time "answer students questions"; the use of social media takes place in time. Here we can observe what Bauman (2011) identifies as puantilistic time (when time gets pressed into one point or a dot) or what McLuhan (2003) speaks of as the change of linearity into concentricity. A research participant mentions the experience of time as if it is disappearing somewhere: "Sometimes you spend hours in front of a computer screen looking for information and after that you cannot understand what you were doing at the computer, as if it eats up your time," as if technologies eat up the time.

The sub-theme of fragmented communication reflects what Bauman (2011) identifies as liquid modernity, the feature of which is fragmentation and the mosaic nature of human existence and of communication. The research participants name it as the disappearance of direct contact: "Direct contact with the students is disappearing in social media"; regarding disrupted communication, "It irritates me when you are talking to a person and he/she is connected to his iPad or something and he/she is constantly on Facebook, it seems he/she is not communicating with you."

Some research participants demonstrate their reservations speaking about privacy in the medium of social media. The sub-theme of cautiousness about privacy includes fear of being followed, insecurity about personal data use, and reluctance to reveal personal data. One research participant claims, "Personally I do not use Facebook because I do not know how my personal data are used," but the concern about personal data is in fact influenced by media where one can find various conspiracy theories fueling cautiousness about social media use. As another research participant says, "Somehow I became afraid of all the networks, because Google search can see what you are doing, all my moves in the digital space are registered." Others indicate that they simply do not feel like sharing their personal information and prefer having their personal space uncovered in social media: "Maybe it's my character, I don't like showing my photographs, telling about myself; it is too personal (Table 3.16)."

In the super-ordinate theme of change manifestation, there are two themes:

- Social-media-generated changes;
- Challenges adapting to change.

The theme of social-media-generated changes contains the following sub-themes:

- A shift to "virtuality," which includes virtual stimulations of real life, moving toward video information, although information is still heavily text based, and moving to e-learning, learning in an online environment instead of in brick-and-mortar institutions (Mason and Rennie 2008).
- Changes in study processes by adopting more open study methods, changing roles—as a lecturer is becoming more of a tutor—and re-planning and changing content.
- The research participants identify that the teacher role is changing: "The study methods are becoming more open to teachers and students" and "The teacher role changes, he/she becomes more a tutor." The change of educator role is envisioned by Kop (2010), who regards educators as "trusted human filters of information." Another research participant indicates that, "Everything is so fast that you realize that you have to change the content as well; for example, if I use tests on Moodle, I know that students will get the results immediately, I'll save time for checking, and then I have to plan what activities I will be doing after the test." This resonates with Mason and Rennie's (2008) statement that, "Simply putting up a course

Table 3.16 Change manifestations (teachers)

Meaning unit	Abstracted unit	Sub-theme 1	Sub-theme 2	Theme
There could be simulations of real-life situations, when learning takes place through virtual experience, living through and creating	Simulations of real-life situations, learning takes place through virtual experience	Virtual simulations of real life	Shift to "virtuality"	Social media generated changes
There appears more video information not only textual	More video information	Moving to video information		
There will be more lectures online, more e-learning	More lectures online, more e-learning	Moving to e-learning		
If a person becomes mobile while learning and can learn not depending on place and time, then there is a question if physical institutions will remain	There is a question if physical institutions will remain			
The study methods are becoming more open to teachers and students	Study methods are becoming more open	Changes in study methods and roles	Changes in study process	
The teacher role changes, he/she becomes more a tutor	Teacher becomes more a tutor			
Everything is so fast that you realize that you have to change the content as well, for example, if I use tests on Moodle, I know that students will get the results immediately, I'll save time for checking, and then I have to plan what activities I will be doing after the test	Everything is so fast that you have to change the content, to plan the activities	Re-planning and changing content		
Teachers need to be trained, they need to know the newest tendencies, someone to be shown, to be explained how technologies work	Teachers need to be trained, they need to know the newest tendencies	Need to know newest tendencies	Constant self-improvement and training	Challenges adapting to change
What is more Moodle environment is renewed constantly, just recently the system has been renewed and now it is necessary again to learn the new format, what has changed, even the structure how to upload the test has changed, the environment constantly changes, develops and you have always to improve yourself	Constant change of the environment causes constant self-improvement	Need for constant self-improvement		

(continued)

Table 3.16 (continued)

Meaning unit	Abstracted unit	Sub-theme 1	Sub-theme 2	Theme
If the media doesn't change and you get used to it and then it changes out of a sudden, It's always a shock	When media changes, a chock is experienced	Shock because of changes		
It depends on the teacher, how he/she uses social media, if he/she explains the students what to do, how to find the information. Students need the explanation	It is necessary to explain the students where to find information	Necessity for consultations both for lecturers and students	Demand for consultations	
It seems to me teachers need individual consultations and students certainly need instructions	Need for individual consultations for teachers			

online without any modification would not work" and Pollock's (2013) argument that what is done in face-to-face classes cannot be simply be transferred online.

The theme of challenges adapting to change includes:

- Constant self-improvement and training;
- Demand for consultations.

The research participants identify a constant need for self-improvement and reveal that sometimes changes are experienced as a shock by them: "If the media doesn't change and you get used to it and then it changes suddenly, it's always a shock." Mason and Rennie (2008) express a revealing insight, identifying that in many cases staff are supposed to educate themselves, which causes additional tension and is usually perceived as additional workload on the staff. It could be the reason why the research participants identify that there is a necessity for consultations, both for the staff and for the students. It really reveals a huge problem that courses developed in discontinued software get lost and the teachers have to rearrange and reprogram everything again which is a great waste of time and effort.

3.2.1.2   Student Experience

Student research participants enrich the research data with their experiences. They point out that they use social media for:

- Communication;
- Collaboration;
- Sharing information;
- Information search (Table 3.17).

The research participants say that they use social media for communicating with their group mates and friends: "I use Facebook for communicating with my group mates… I use social media to communicate with friends, colleagues… I studied abroad before, so to communicate with my ex-group mates." They also use social

Table 3.17 Social media application (student use of social media)

Meaning unit	Abstracted unit	Sub-theme 1	Sub-theme 2	Theme
I use Facebook for communicating with my group mates	Use social media for communicating with my group mates	Communication with group mates	Communication	Student use of social media
I use social media to communicate with friends, colleagues. I studied abroad before, so to communicate with my ex-group mates	Use social media to communicate with friends, colleagues	Communication with friends		
Social networks are for collaboration, there are Google forms where one document could be prepared by many people	Social networks are for collaboration, documents could be prepared by many people	Collaboration for preparing documents	Collaboration	
I participate in group discussions and try to find out the answer to the questions together	Participate in group discussions, try to find answers to the questions	Collaboration looking for answers		
I use Facebook to exchange of information with friends and acquaintances	Use Facebook to exchange information with friends	Information exchange	Sharing information	
It's important for me to take photographs and to upload information on social media because I know that that there are people in my organization who are following my life and that I am a role model for them	Upload information on social media because I know there are people who are following	Share information for people to follow		
As well we have Facebook group where we exchange information to find out about homework about lectures and so on	We exchange information to find out about homework, lectures, etc.	Sharing information on studies		
As I am a monitor of my group so I disseminate information for my group mates, participate in group discussions	I disseminate information for my group mates			
Speaking about Moodle we can find all the information necessary for our studies and sometimes we search the Internet for information	We can find the information necessary	Finding necessary information	Information search	
I use social media for information search	Use social media for information search			
Mainly I follow the events because physically you cannot run around all the events in all the pages and here everything is in one place	I follow the events, everything is in one place			

media for collaboration while preparing documents and looking for answers together: "Social networks are for collaboration, there are Google forms where one document could be prepared by many people… I participate in group discussions and try to find out the answer to the questions together."

Students share information with friends, upload information so that others can follow their activities and also share with their group mates the information connected to their studies: "I use Facebook to exchange of information with friends and acquaintances… It's important for me to take photos and to upload information on social media because I know that that there are people in my organization who are following my life and that I am a role model for them… As well we have a Facebook group where we exchange information to find out about homework about lectures and so on."

The research participants also state that they use social media for information searches connected to their studies and various events and for any other necessary information: "I use social media for information search… Mainly I follow the events because physically you cannot run around all the events in all the pages and here everything is in one place." The research participants experience the availability of information in one place (Table 3.18).

Student-perceived advantages include the sub-themes of:

- Compact information;
- Availability of information;
- Presence indication;
- Collaboration possibilities.

Students acknowledge that information is in one place—usually on their technological devices—and as well they take photographs of the learning material presented by the lecturers; in this way, technology makes keeping information really comfortable.

Another advantage is the availability of information, as the research participants say, "Information is available all the time, if I don't go to the lectures, and I find everything in our Facebook group." The next point indicated by the research participants is presence indication: "I like writing messages on Facebook because you can see immediately who is connected and if he/she has seen the message, then communication seems to be livelier." Pollock (2013) writes about immediate or "synchronous" presence in distance communication, which increases interaction among students. Also, according to the theory of social presence, the effectiveness of communication depends on the level of social presence (Sallnas, Rassmus-Grohn, and Sjostrom 2000). Students also indicate collaboration possibilities: "It's very comfortable, we have a group on Facebook where we communicate and discuss with group mates, share study material, help each other (Table 3.19)."

Student-perceived issues while using social media include:

- Time consumption;
- Distraction;
- Aggravating content;

Table 3.18 Perceived advantages (students)

Meaning unit	Abstracted unit	Sub-theme 1	Sub-theme 2	Theme
I don't use paper resources, I don't carry, everything is in my iPad	I don't use paper resources, everything is on my iPad	Information in one place	Compact information	Student-perceived advantages of social media
I take photographs of learning materials, of teachers' presentations	Take photographs of learning materials	Compact form of learning materials		
Information is available all the time, if I don't go to the lectures, I find everything in our Facebook group	Information is available all the time	Easy access to information in time and space	Availability of information	
On one hand it saves time because you can find information quickly	I can find information quickly			
It makes your life easier because you can find information easily and you save time	You can find information easily			
You can find everything being at home, you don't need to go to the library, to search plenty of books	You can find information being at home			
Actually you have your group of friends in your page and you can see if the person is connected or not and it helps you	You can see if a person is connected or not and it helps you	Helpful to see presence status	Presence indication	
I like writing messages on Facebook because you can see immediately who is connected and if he/she has seen the message, then communication seems to be livelier	You can see who is connected and communication seems livelier	Presence online animates communication		
It's very comfortable we have a group on Facebook where we communicate and discuss with group mates share study material, help each other	We communicate and discuss with group mates, share study material help each other	Possibility of collaborative study	Collaboration possibilities	

- Fragmentation of communication;
- Difficulties processing information.

The research participants identify a tendency to spend more time online, and in such a way social media causes time consumption for them. The two sub-themes of time consumption and distraction appear to be closely related because the research participants disclose experiencing loss of time because of slipping into the social media world, reading articles, watching videos, or communicating with friends: "On the other hand, you lose time because when you start looking for what you need, somehow you slip into a social network or start looking at a different topic, there are

Table 3.19 Perceived problems (student issues using social media)

Meaning unit	Abstracted unit	Sub-theme 1	Sub-theme 2	Theme
You spend more time at the computer, sometimes you lose time, start watching videos or communicating with friends	Spend more time at the computer, sometimes lose time	Increased time online	Time consumption	Student-perceived issues of social media
On the other hand, you lose time because when you start looking for what you need, somehow you slip into social network or start looking at a different topic, there are many distractions	Somehow you slip into social network and start looking at a different topic	Getting distracted by social media	Distraction	
You come home and first you connect to Facebook and start communicating, watching videos, reading articles, and this distracts your attention and it's difficult to disconnect	You start communicating, watching videos, reading articles, and this distracts your attention and it's difficult to disconnect			
When I read the news, I find mostly bad news like explosions, murders and it influences my mood	Find mostly bad news which influence my mood	Upsetting content	Aggravating content	
Communication on social media doesn't fell natural for me because people create their profiles, certain images and when you communicate, you communicate with created images	Communication on social media doesn't feel natural, you communicate with created images	Alienated communication	Fragmentation of communication	
More and more often I meet people who are constantly connected to Facebook and even when you are with them, they still are busy on their phones and communication is disrupted	When you are face-to-face with people they still are busy on their phones and communication is disrupted	Disrupted face-to-face communication		
There is such an abundance of information that sometimes it's difficult to choose the appropriate information and sometimes you feel lost	There is such an abundance of information that sometimes it's difficult to choose the appropriate information	Abundance of information	Difficulties processing information	
Social media is a good thing but there is a lot of rubbish, unreliable information	There is a lot of rubbish, unreliable information	Unreliability of information		

many distractions… start watching videos or communicating with friends." As the research participants indicate, usually people have the intention to check their social network account just for five minutes, but these five minutes grow into 20 min or later into 40 min.

Student distraction by social media has been discussed by the researchers. The Kaiser Family Foundation carried out a survey in 2010 and reported that children ages eight to 18 spend approximately seven or more hours per day using social media networks for entertainment (Kaiser Family Foundation, 2010). Cisco research (2011) reveals that 1 in 3 college students consider the Internet to be just as fundamental as air, water, shelter, and food are. It seems that the Internet connection is becoming perceived as vital by the majority. McCoy (2013), in his research, observes that the top reasons for using digital devices for non-class purposes, according to students, are staying connected (70%), fighting boredom (55%), and doing related class work (49%). The most commonly acknowledged disadvantages were that they don't pay attention (90%) or miss instruction (80%).

The research participants indicate that they experience distraction during their studies at home as well. The sub-theme "aggravating content" reveals the research participant experience of being affected by the content found on social media: "When I read the news, I find mostly bad news like explosions, murders, and it influences my mood." Bandura (2002) proved in his research on the influence of audiovisual content on the content users that social media content has the power to affect the users.

The sub-theme of fragmented communication includes the issues of alienation of communication when communication occurs among the created images on social media—"Communication on social media doesn't feel natural for me because people create their profiles, certain images, and when you communicate, you communicate with created images"—and disrupted face-to-face communication—"More and more often I meet people who are constantly connected to Facebook and even when you are with them, they still are busy on their phones and communication is disrupted"—rather than in person.

The sub-theme of difficulties of processing information is related to the abundance of information, when it becomes difficult to choose what is appropriate while gliding across the ocean of information—"There is such an abundance of information that sometimes it's difficult to choose the appropriate information and sometimes you feel lost"—and dealing with the unreliability of information when the user has to check the sources—"Social media is a good thing but there is a lot of rubbish, unreliable information (Table 3.20)."

In the super-ordinate theme of change manifestations, clearly standing out is the theme of students' needs and requests, which includes:

- Need for manageable media environment;
- Request for teacher development;
- Request to acknowledge multitasking;
- Request for open teacher–student communication;
- Need for media literacies;

Table 3.20 Change manifestations (students)

Meaning unit	Abstracted unit	Sub-theme 1	Sub-theme 2	Theme
I noticed that during the day I need approximately 5 rubrics but every time I need to search for them because they are hidden somewhere, the system should be user-friendly not so complicated especially for the beginning students	Every time I need to search for rubrics because they are hidden, the system should be user-friendly	Need for user-friendly system	Need for manageable media environment	Students' needs and requests
Speaking about university Moodle system, it should be user-friendlier, simpler, maybe simpler design	It should be user-friendlier, simpler, maybe simpler design			
I think that people aren't interested in innovations, so there should be more education and training on modern technologies	There should be more education and training on modern technologies	Need for training on modern technologies	Request for teacher development	
I showed my teacher my digital reader where everything is in one place, dictionary and everything and she didn't know such a thing, there should be more information, more training about the new forms of social media	There should be more training about the new forms of social media	Need for training		
Teachers try to help, they explain all the systems but sometimes there are teachers maybe not created for modern technologies, they don't know how to use social media and it makes the study process more difficult	There are teachers who don't know how to use social media			
Some teachers are maybe old-fashioned, they don't try innovations	Some teachers don't try innovations	Need to apply innovations		
If social media was used actively in university public life it would be useful both for students and teachers, would make the study process easier like when I was studying in Denmark, there was a page on Facebook through which you could access study resources	If social media was used actively in university public life it would be useful both for students and teachers, would make the study process easier	Active use of social media in public university life		

(continued)

Table 3.20 (continued)

Meaning unit	Abstracted unit	Sub-theme 1	Sub-theme 2	Theme
I like working in the background of Facebook, from time to time I look at it for a few minutes and then return back. People complain that the attention gets scattered but if I didn't look at Facebook, I might look through the window or might talk with someone	Like working with Facebook in background	Studying with Facebook in background	Request to acknowledge multitasking/task-switching	
There are a few teachers on Facebook and I like it, I like when I pass someone at university and I know the person from social media	Like knowing teachers from social media	Wish for informal information from social media	Request for open teacher–student communication	
We have one teacher who you can ask on Facebook doesn't matter at four o'clock in the morning "Sorry professor, I don't understand" and he will answer and explain but others are old-fashioned, they stress the importance of their personal space. Maybe if their attitude changed we could communicate more easily	Teachers stress the importance of their personal space. Maybe if their attitude changed we could communicate more easily	Need for easier communication with teachers		
We need critical thinking because there is a lot of information and not everywhere it is so reliable, you need to sort it out	We need critical thinking for sorting out information	Need for strategies to deal with information	Need for media literacies	
There is a lot of information, so sometimes it's difficult to select the appropriate one, sometimes you might feel scared or lost	Difficult to select the appropriate information			
When there is a possibility to find the information easily and teachers know this, they give us too much extra work, so sometimes I think it would be better if social media stopped improving	Teachers know that information could be found easily and they give us too much extra work	Too much extra work	Need for managing study load	

- Need for managing study load.

The research participants state that when using university pages and university social media environments, they face inhibitions as the users because of the complexity of the environment. They express a wish for a more user-friendly environment: "I noticed that during the day I need approximately five rubrics but every time I need to search for them because they are hidden somewhere, the system should be user-friendly, not so complicated, especially for the beginning students… Speaking about university Moodle system, it should be user-friendlier, simpler, maybe simpler design." It appears that students experience the complexity of the environments that I as a researcher experienced as well, when on the opening page you just stare for a while trying to orient yourself regarding the functions, sometimes digging for them helplessly until you realize you have to approach a colleague or a system administrator for a consultation.

The following three sub-themes—request for teacher development, request to acknowledge multitasking and request for an open teacher–student communication—are closely connected to the issue of the pedagogical relationship discussed in a later section. The research participants state that sometimes they face the reality that some teachers are not acquainted with technologies and sometimes it causes difficulties: "Teachers try to help, they explain all the systems but sometimes there are teachers maybe not created for modern technologies, they don't know how to use social media and it makes the study process more difficult… Some teachers are maybe old-fashioned, they don't try innovations." This shows that there is a need for a constant teacher training since technologies change and one way of adapting to change is training and learning how to use new technologies and new types of social media in the study process.

Students also express their wish for acknowledging multitasking. As discussed above, research on multitasking has proved to be contradictory, with some researchers advocating for it but many researchers finding it irrelevant and damaging students' grades (Fewkes and McCabe 2012). Not everything is so simple, however; as McCoy (2013) observes, 55% of the respondents in his survey used digital devices for multitasking to fight boredom. Similarly, a research participant states: "I like working in the background of Facebook, from time to time I look at it for a few minutes and then return back. People complain that the attention gets scattered but if I didn't look at Facebook, I might look through the window or might talk with someone." Also, Rouis, Limayem, and Salehi-Sangari (2011), in their research, proved that students with high self-regulation do not experience any effects on their marks caused by multitasking. Kirschner and De Bruyckere (2017) claim that multitasking is not possible due to the cognitive architecture of human beings and introduce task-switching which may relate to the observations by McCoy (2013) of student digital device use to fight boredom have a short break. Here one of the authors may recall visiting her colleague in the office environment where the colleague's computer would flash a reminder of the need for a break every 15 min. The colleague explained that it was her self-regulative intention to have short breaks in her intensive work schedule. It seems that engaging activities and acknowledging student right for self-regulation on task-shifting as inherent to some students might be the answer to the issue.

The main feature of the pedagogical relationship is open and caring educator–student communication (van Manen 2013), and likewise the research participants indicate their longing and need for an open teacher–student communication, including the sub-themes of a wish for informal information from social media and a request for easier communication with teachers. As the research participants put it: "There are a few teachers on Facebook and I like it, I like when I pass someone at university and I know the person from social media… We have one teacher who you can ask on Facebook doesn't matter at four o'clock in the morning, 'Sorry professor, I don't understand' and he will answer and explain, but others are old-fashioned, they stress the importance of their personal space. Maybe if their attitude changed we could communicate more easily." It is obvious that students value teacher approachability and the ability to keep informal communication. It also reveals the advantageous use of asynchronous mode of social media when a student may ask a question "at four a.m." and receive the answer later during the day.

Something else that stands out is the necessity for media "literacies" identified by Rheingold (2010), which include: attention, participation, collaboration, network awareness, and critical consumption. The research participants identify that they need strategies to deal with the information flow and the ability to choose the necessary information for them, so they need a critical consumption strategy most: "We need critical thinking because there is a lot of information and not everywhere it is so reliable, you need to sort it out."

The sub-theme of the need for managing the study load reveals the fact that sometimes students experience overload, though the credit and the study hour system seems to be well-documented even in the international EU framework documents. As one research participant states: "When there is a possibility to find the information easily and teachers know this, they give us too much extra work, so sometimes I think it would be better if social media stopped improving." The research participant even expresses a wish for social media to stop improving, but here what is really experienced is the need for managing study load, which, according to researchers, is closely related to student ability to manage time within their coursework (Walker and Siebert 1990). This indicates, however, that students need some kind of guidance or educational consultations on how manage their study load.

3.2.1.3 Administrator Experience

Administrator research participants identify their social media uses. The theme of administrator use of social media contains the sub-themes of:

- System maintenance;
- Providing trainings;
- Sharing information;
- Communication;

- Information source;
- Advertising their institution (Table 3.21).

First, administrators maintain the LMS by administering the system, developing it, and integrating social media into the system so that everything is in one place. The administrator group works with all university students and teachers: "We are a group who work with approximately 17000 students and actually all the teachers; we are administering, tutoring and monitoring all the university in e-learning environments."

Table 3.21 Social media application (administrator use of social media)

Meaning unit	Abstracted unit	Sub-theme 1	Sub-theme 2	Theme
We are a group who work with approximately 17,000 students and actually all the teachers, we are administering, tutoring and monitoring all the university in e-learning environments	We are administering, tutoring and monitoring all the university in e-learning environments	System administration	System maintenance	Administrator use of social media
We are improving the system, developing it	Improving the system, developing it	System development		
The main study system is Moodle, some teachers use it in a mixed way with Facebook and Twitter but Moodle provides an opportunity to integrate everything and use all social media from one place	Moodle allows to integrate everything and use all social media from one place	Integration of social media into the system		
Our department provides trainings focusing on how practically teachers can work with social media	Provide trainings on social media use	Trainings on social media use	Providing trainings	
We even teach others	Teach others			
I use media to transfer information	Use to transfer information	Information transfer	Sharing information	
I upload photographs on social media	Upload photographs	Share photographs		
I use social media for communication, meeting friends, finding the old ones	Use social media for communication, meeting friends	Communication with friends	Communication	
We use social media for communicating with colleagues	Use for communicating with colleagues	Communication with colleagues		
As well I use for hobbies and finding out about the events	Use for hobbies and finding out about the events	Information about events	Information source	

(continued)

Table 3.21 (continued)

Meaning unit	Abstracted unit	Sub-theme 1	Sub-theme 2	Theme
I use social media for my hobbies	Use social media for my hobbies	Information on hobbies		
Our institution, we have a page for advertising, for making our institution visible in public	We have a page for making institution visible in public	Making institution visible in public	Advertising institution	
We use for dissemination of information for our institution, attracting students	Use for dissemination of information and attracting students	Attracting students		
We publish information about our distant studies, timetables, what subjects are available	Publish information about distant studies	Promoting distant studies		

Although the official university-maintained LMS is Moodle, administrator group also integrates social media into the system: "The main study system is Moodle, some teachers use it in a mixed way with Facebook and Twitter but Moodle provides an opportunity to integrate everything and use all social media from one place."

The group also provides trainings—"Our department provides trainings focusing on how practically teachers can work with social media"—and advertises the institution by making the institution visible to the public, attracting students and promoting distant studies—"We use for dissemination of information for our institution, attracting students… We publish information about our distant studies, timetables, what subjects are available."

As well, administrators use social media like the other groups of the research participants, for sharing information, communication and as the source of information. The research participants state that they transfer information, upload photographs on social media, communicate with colleagues and friends, and find out about the events and useful information for their hobbies (Table 3.22).

The theme of administrator-perceived advantages of social media includes:

- Availability of information;
- Overall impact as advanced institution;
- Increased information exchange.

The research participants observe that information remains permanently—"Increased information exchange"—and almost any answers to any questions are available on Google: "I liked the quote which I found that in 2011 there were billions of questions searched on Google, so then there is a question who would people address these questions to before Google." The research participant reveals the experience of information availability in a very precise and accurate way by simply posing the question of who would people address these questions to before Google, and, as I myself remember the times before Google, I feel that then traditional information

Table 3.22 Perceived advantages (administrators)

Meaning unit	Abstracted unit	Sub-theme 1	Sub-theme 2	Theme
Increased information exchange	It is enough to create something once, you don't have to upload every time	Information remains permanently	Availability of information	Administrator-perceived advantages of social media
I liked the quote which I found "In 2011 there were billions of questions searched on Google" so then there is a question who would people address these questions before Google	There is a question who would answer questions before Google	Available answers on Google		
It's very convenient you don't need to carry any material, everything is in social media environment	Everything is in social media environment	Availability of resources		
When our university started using Moodle, the general level of Moodle use increased in the whole country	Our university using Moodle increased the general level in the whole country	Institutional social media use effects	Overall impact as advanced institution	
Lithuania is a small country and our staff work in other universities and colleges, so it was like a chain reaction, our institution raised the level at the same time raising the level of other institutions	It was like a chain reaction, our institution raised the level at the same time raising the level of other institutions	Impacting other institutions		
It's easier to communicate with people	Easier to communicate with people	Easier communication	Increased information exchange	
It's quicker to share the information, you just upload it and everyone can see it	Quicker to share information			

resources were used, people would consult available books or experts, but still many questions would remain unanswered; nowadays, information literary is at the tips of your fingers, and sometimes it is enough to just to push a button or touch the screen.

The sub-theme of overall advanced impact of the institution reveals that implementation development and application of social media in one institution affects the environment and other parts of the educational space. As one research participant says, "When our university started using Moodle, the general level of Moodle use increased in the whole country... Lithuania is a small country and our staff work in other universities and colleges, so it was like a chain reaction, our institution raised the level at the same time raising the level of other institutions." The process is experienced by the research participant as "a chain reaction," a process wherein application of technology in one educational institution influences other institutions. The sub-theme of increased information exchange reveals that social media makes it easier to communicate with people and quicker to share information: "It's quicker to share the information, you just upload it and everyone can see it (Table 3.23)."

Administrator-perceived issues contain the following sub-themes:

- Need for increased communication;
- Demand for staff development;
- Need to make training attractive;
- Censorship of students' opinions;
- Time consumption;
- Knowledge commoditization;
- Distraction.

The sub-theme of increased communication reveals administrator experience with students expressing their request for more communication between teachers and students and, as well, more communication across subjects. As will be discussed in a later section, analyzing the pedagogical relationship, it appears that students' wish for more informal and more extensive communication with teachers is a natural student longing for the pedagogical relationship in which they would be allowed to learn and grow. As the research participants in the administrator group observe, "Our last discussions with the students showed that they lack communication with the teachers... Teachers who use Facebook are admired and highly evaluated by students, they say, 'He's God'... Students define it as an additional channel with the teacher." The importance of the pedagogical relationship and of teacher communication and attention to students is genuinely expressed by the phrase "He's God," indicating that the one who establishes the right pedagogical relationship and communicates with students in multiple channels, allowing them to learn and develop as personalities, is valued like "God," the one who gives everything—or, according to religious beliefs, the one who gives the kingdom of heaven.

Another request by students, as experienced by administrators, is a possibility to communicate across subjects, not only in closed groups within the limits of subject matter; it reveals how much communication is important for students. Administrators also indicate a demand for staff development, which includes a training necessity and the need to develop acceptance of social media, to adapt a social media teaching style, to balance content and tools, and to adapt to being recorded. In fact, it reveals how challenging social media use for the teachers is and how much training is needed. It is not enough just to learn technical skills to operate social media tools but it is also necessary to balance teaching and tools. Similarly, Bates (2005) argues that there is the need to redesign and reorganize teaching in order to successfully and fully apply the new technology, which—according to the author—is often ignored. The next

Table 3.23 Perceived problems (administrator issues of using social media)

Meaning unit	Abstracted unit	Sub-theme 1	Sub-theme 2	Theme
Our last discussions with the students showed that they lack communication with the teachers	Discussion showed the lack of student–teacher communication	Need for more student–teacher communication	Need for increased communication	Administrator-perceived issues
Teachers who use Facebook are admired and highly evaluated by students, they say, "He's God"	Students highly evaluate teachers using Facebook			
Students define it as an additional channel with the teacher	Additional channel with the teacher			
There is a desire expressed by students to enable broader communication among everybody because now they can only communicate within the subjects, they would like to communicate across subjects	Desire expressed by students to communicate across subjects	Enabled communication across subjects		
We have everything, just we need to learn to use social media, I think training will help	We need to learn to use social media	Training necessity	Demand for staff development	
There is some reluctance, there are declarations of some study fields claiming that there shouldn't be social media only face-to-face teaching, maybe persuasion can help, just demonstration and positive examples	Persuasion and demonstration may help to accept social media	Need to develop acceptance of social media		
Sometimes teachers want to have as much social media as possible but forget about the content of the subject that different content needs different approach though the used tools are the same but the approach should be different	Sometimes teachers want to use social media and forget about the content	Need to balance content and tools		
Teachers are more constrained in front of the camera if we are speaking about recording lectures	Teachers are more constrained in front of the camera	Need to adapt to being recorded		
The style of teaching changes, the teacher becomes more a moderator, especially in social media environment he/she is even called a moderator	The style of teaching changes in social media	Need to adapt social media teaching style		

(continued)

Table 3.23 (continued)

Meaning unit	Abstracted unit	Sub-theme 1	Sub-theme 2	Theme
Earlier training courses for teachers how to use social media were popular, now few people come, the initiative is fading, and maybe the training appeared uninteresting	Earlier training courses were popular, the initiative is fading	Decreasing popularity of training courses	Need to make training attractive	
Information in social media is available to everyone and more difficult to control, more difficult to control students' opinions and responses because if they are negative, they shouldn't be publicized	It's difficult to control students opinions in order not to publicize the negative ones	Wish to control students' opinions	Censorship of students' opinions	
It's is a challenge to start using social media, it takes time	A challenge to start using social media, takes time	Time for mastering social media	Time consumption	
In open databases you can get knowledge free of charge but if you want to get a certified credit you have to pay, so it's relatively free of charge	In open databases you get knowledge but if you want to get a certified credit you have to pay	Learning versus paid certification	Knowledge commoditization	
I get distracted even in the working environment if I see that someone connected or disconnected, it distracts me	I get distracted seeing someone connected or disconnected	Getting distracted	Distraction	

sub-theme, also connected to training, reveals administrator experience that there exists a need to make training attractive: "Earlier training courses for lecturers on how to use social media were popular, now few people come, the initiative is fading, and maybe the training appeared uninteresting."

The sub-theme of censorship of students' opinions may be natural for institutions; however, it goes in tune with Foucault's (1998) insights on institutional practices to discipline and control. A research participant reveals: "Information in social media is available to everyone and more difficult to control, more difficult to control students' opinions and responses because if they are negative they shouldn't be publicized." It is natural for an institution to try to keep a positive image, but here we face a desire to censor students' opinions by making sure that negative opinions are not publicized.

The sub-theme of time consumption is mentioned by all three groups of research participants. The administrator group acknowledges that adopting and using social media takes time: "It's is a challenge to start using social media, it takes time." Also, administrators mention knowledge commoditization: "In open databases you can get knowledge free of charge but if you want to get a certified credit you have to pay, so it's relatively free of charge." At first sight, it seems that social media enables the learner to make more decisions on learning Siemens' (2004) "connectivism" theory; however, Mason and Rennie (2008) seem to be right, indicating "knowledge commoditization" and "the student as customer" paying for certification or credits.

The sub-theme of distraction is also mentioned by all three groups of research participants. The administrators observe that social media acts as a distraction in their daily work: "I get distracted even in the working environment; if I see that someone connected or disconnected, it distracts me (Table 3.24)."

The theme of institution support contains:

- Infrastructure development;
- Institutional promotion of social media;

Table 3.24 Change manifestations (administrators)

Meaning unit	Abstracted unit	Sub-theme 1	Sub-theme 2	Theme
Our institution supports social media use in studies, we create all the infrastructure for all the faculties, we try to improve, to add new tools according to the needs	We create all the infrastructure, we try to improve, add new tools	Creating infrastructure	Infrastructure development	Institution support
We are redefining teacher working places, adding new equipment, including possibilities not working onsite, our regulations allow moving lectures into social media environment	Redefining teacher working places, moving lectures into social media environment	Redefinition of working places		
The initiative to use social media comes from our marketing department not from the teachers, the marketing department is oriented toward student needs, to what other universities are using, to the whole environment	The initiative to use social media comes from marketing department	Marketing department initiative for social media use	Institutional promotion of social media	
Social media is the priority of our institution not only lecturer and student contact at the work place but also their contact beyond the limits of the classroom	Social media is the priority in our institution	Priority in institution		
Like with the application of Moodle in our institution, there was a system of promotion, through finances and training and now we have a big progress even without promotion	There was a system of promotion through finances and training	System of promotion through training	Staff support	
Our department provides training courses for our lecturers and as well we provide everyday help if there are any questions	Our department provides training and everyday help	Training and consultations provided		

- Staff support.

The administrators indicate that the institution infrastructure is created, maintained, and developed: "Our institution supports social media use in studies, we create the infrastructure for all the faculties, we try to improve, to add new tools according to the needs." They also identify that lecturer working places are being redefined, including possibilities of not working onsite and moving lectures into the environment of social media. At the same time, it is pointed out that the initiative to use social media as a priority comes from the institution: "The initiative to use social media comes from our marketing department, not from the lecturers; the marketing department is oriented towards student needs, towards what other universities are using, towards the whole environment." So in fact, the institution supports social media use, creating the infrastructure, having it as priority and providing staff support: "Our department provides training courses for our lecturers and as well we provide everyday help if there are any questions (Table 3.25)."

There also stands out the theme of administrator caution regarding social media use in university study processes, which shows that there are some reservations on behalf of administrators. They include:

- Caution about privacy;
- Caution about security;
- Critical view on content;
- Preference for face-to-face communication;
- Caution about transience of social media;
- Imposed age limitations.

The administrator experience reveals that they are cautious about their privacy and express their reluctance to publicize their private information: "I have a position not to publicize my personal information, my contacts are on the Internet and this is enough." A similar issue is noted by Gross and Acquisti (2005) that social networks demonstrate the characteristic of being more levelled and information is available to more friends, but then trust might decrease within online social networks as the subject is not sure about the broad accessibility of personal information.

The sub-theme "caution about security" is related to the objectively existing circumstances that, in some cases, due to security issues, social media cannot be used or is not used: "Sometimes social media is not used objectively; for example, in the faculty of public security no social media is used because of security issues." In such cases, only closed systems ensuring security could be used.

Some administrators express their critical position about social media content: "I don't participate in Facebook and other social networks because it's a waste of time uploading or watching daily photos of other people. I don't understand publicizing your personal life; such writing like 'Today I have a headache, could you give me a piece of advice' seems silly to me." Smith (2009), however, indicates that social media opened up the possibility to express opinions that may not be high-quality content, expectations of which are expressed in administrator reservation; but, according to Smith (2009), opinions could be used for building up improvement

Table 3.25 Reservations (administrators)

Meaning unit	Abstracted unit	Sub-theme 1	Sub-theme 2	Theme
I have a position not to publicize my personal information, my contacts are on the Internet and this is enough	I have a position not to publicize my personal information	Reluctance to publicize private information	Caution about privacy	Caution regarding social media use
Sometimes social media is not used objectively, for example, in the faculty of public security no social media is used because of security issues	In the faculty of public security, no social media is used because of security	Security limits social media use	Caution about security	
I don't participate in Facebook and other social networks because it's a waste of time uploading or watching daily photographs of other people. I don't understand publicizing your personal life; such writing like "Today I have a headache, could you give me a piece of advice" seems silly to me	It's a waste of time uploading and watching daily photographs of other people	Unattractive daily life content	Critical view on content	
Face-to-face communication isn't going to be substituted by anything; for example, parents with disabled children come to us to inquire about the studies and we say, "No problem, they can study distantly" but the parents say, "We don't want to isolate our child; we want him to be with his peers"	Face-to-face communication isn't going to be substituted by anything	Stressing the importance of face-to-face communication	Preference of face-to-face communication	
We shouldn't forget that social media is in the air, in the cloud, and if one day a thunder hits this cloud, we could remain without anything, so I'm reserved; for me the cloud is the cloud, technology is good but safety first	Social media is in the cloud, I'm reserved, for me the cloud is the cloud	Doubts about social media durability	Caution about transience of social media	
If we look at the average age of our teachers and the foreign language they can use, we will understand that English is not their strength and the newest social media applications are in English	If we look at our teachers age we can understand that English is not their strength	Teacher age connected to mastery of English	Imposed age limitations	
There are old teachers who are used to traditional teaching; students want new means, so a kind of breakthrough is necessary	There are old teachers used to traditional teaching	Teacher age connected to traditional teaching		

and development. Agichtain et al. (2008) assert that user-generated content in social media varies drastically—it contains both high-quality content and low content.

Administrators also express the existing preference for face-to-face communication. Their experience demonstrates that there are certain groups of people who prefer face-to-face interaction: "Face-to-face communication isn't going to be substituted by anything; for example, parents with disabled children come to us to inquire about the studies and we say, 'No problem, they can study distantly' but the parents say, 'We don't want to isolate our child; we want him to be with his peers'." The example reveals that social media interaction could be considered an additional channel to face-to-face interaction.

Another cautious attitude is related to the transience of social media. Administrators experience the seeming non-durability of social media like "being in the air." As one research participants points out, "We shouldn't forget that social media is in the air, in the cloud, and if one day a thunder hits this cloud, we could remain without anything, so I'm reserved; for me the cloud is the cloud, technology is good but safety first." Also, administrators report the imposed age limitations, claiming that teacher age is related to reluctance of social media use, prevalence of traditional teaching methods, and poor knowledge of English, which all might be impediments to the application of social media.

The research participants indicate that mastery of English is an important factor in adopting social media use: "If we look at the average age of our teachers and the foreign language they can use, we will understand that English is not their strength and the newest social media applications are in English." They point out age as well: "There are old teachers who are used to traditional teaching; students want new means, so a kind of breakthrough is necessary." Though it should be admitted that according to UTAUT—the Unified Theory of Acceptance and Use of Technology (Venkatesh et al. 2003)—technology acceptance behavior depends on the age factor, showing better results for younger ages, however; the more weighty factors are the perception of utility and the perception of ease of use. Also, researchers admit that research on job-related attitudes (Hall and Mansfield 1995) reveals that younger workers may consider extrinsic rewards more important and that could be a drive toward enhanced technology acceptance.

3.2.2 Key Emerging Contradictions

According to Arnold (2003), contradictions—the existence of two opposing binaries that cannot be deciphered—are characteristic of technologies. Arnold (2003) identifies three main features for the Janus-faced metaphor that he uses to reveal the irony and paradox of technological conditions:

- The performance of the socio-technical system inherently contains multiple implications that create opposing binaries of are contrasting conclusions.
- These contrasting binaries appear within the same analytical approach.

- These binaries are not erroneous or need any resolution; they are simply co-dependent, intrinsic to the system. Just like deity of Janus in a famous Janusian myth, technology is looking in both directions at the same time; for example, at the same time entering new realms and leaving the previous existence.

The key emerging issues appeared similar to those which were identified in the international inductive qualitative research. First, all three groups of participants mention the difficulties processing information, reliability, and the time spent on social media. Teachers identified such problems as scattered information, advertisement intrusion, material loss, and distraction by unused functions and the necessity of dealing with a non-stop information flow. Students also mentioned the abundance of information, occasional unreliability of information leading to difficulty processing it or a feeling of getting lost in that fragmented information flow: "There is such an abundance of information that sometimes it's difficult to choose the appropriate information and sometimes you feel lost." They also speak about getting distracted by social media and about time used to sort out and obtain the necessary information. Students express their need for critical thinking and for the ability to identify the appropriate information. Administrators also state that time is necessary for mastering social media. All three groups of research participants actually speak about media "literacies" identified by Rheingold (2010), which include: attention, participation, collaboration, network awareness, and critical consumption. The latter stands out as the most important as the research participants identify that they need strategies to deal with the information flow and to choose the necessary information for them.

Another issue is the constantly ongoing contradiction between a more democratic, scattered way of communication on social media and institutional inclination to control. Teachers speak about how social media enables them to feel in control by observing students' activities on social media: "I control… I see what tasks they did, how much they did and what results they got, how many times they attempted the tasks." Administrators feel the inclination to control the opinions expressed by the students: "Information in social media is available to everyone and more difficult to control, more difficult to control students' opinions and responses because if they are negative they shouldn't be publicized." It seems that people get so embedded in their institutional roles that they start performing the controlling roles without questioning whether there could be any possibilities of dealing with different learning options or different opinions.

Meanwhile, students express their wish for more informal, approachable modes of communication with teachers: "We have one teacher who you can ask on Facebook doesn't matter at four o'clock in the morning, 'Sorry professor, I don't understand' and he will answer and explain, but others are old-fashioned, they stress the importance of their personal space. Maybe if their attitude changed we could communicate more easily." Students also feel that their self-regulatory task-shifting, getting into social media from time to time during the lectures, should be acknowledged in its own right. It appears to be a pedagogical challenge as self-regulation levels of the students vary and it is a complex task to cater for all levels (Rouis, Limayem, and Salehi-Sangari 2011). This is especially noteworthy given that researchers Rouis,

Limayem, and Salehi-Sangari (2011) in their research proved that students with high self-regulation do not experience any effects on their marks caused by multitasking (task-shifting); however, observations by Foucault (1998) that educational institutions have ritualized their practice regarding discipline and control seem to persist although the application of social media acts as a driving force toward democratization of educational processes where idealistically students could be catered according to their needs.

Another issue vividly expressed by a student research participant is fragmentation of identity: "You create a certain Facebook profile but you just create a certain image which is not your real self, and when you communicate with a person on Facebook you do not communicate with the person directly, you communicate with a created image, it seems as if the created images communicate among themselves." It resonates with the idea expressed by Bauman (2011) that personal identity becomes an object to be hidden or not expressed, lost in the mosaic game of media stimulated and forced on play of identities where constant identity change, dismissal of the old one and seeking for new manifestations, becomes a must. There is not unambiguous perception, as lecturer research participants observe, the language used by social media participants, their likes and dislikes, activities and even photographs reveal a lot of personality that might not have been expressed elsewhere. What one sees often depends on how one looks (Saevi 2012).

The privacy issue appears to be an additional concern while using social media. Two groups of research participants—lecturers and administrators—express their cautiousness about privacy and a certain unwillingness to reveal personal information, sometimes leading to no or very limited use of social media: "I do not know how my personal data are used," which is reasonable bearing in mind various security issues discussed in the media itself. Student research participants expect more open communication, however, which becomes burdened by fears of breached security. Additionally, some research participants would like to see what their children are doing on social media. In such a case, curiosity wins over privacy concerns: "Now I want to reopen my Facebook account as I would like to see what my son is doing, what his friends are." It resonates with the Momus myth analyzed by van Manen (2010), who observed that Momus became the first to express the desire to access what was hidden in the human heart by means of a technology of surveillance.

Another issue closely connected to the pedagogical relationship analyzed below stands out as the change of teaching/learning processes. Lecturer research participants themselves admit that the lecturer role is changing: "The study methods are becoming more open to lecturers and students" and "The lecturer role changes, he/she becomes more a tutor"; however, there still exists a wish to thoroughly control. Student research participants also express the need for lecturer development: "Lecturers try to help, they explain all the systems but sometimes there are lecturers maybe not created for modern technologies, they don't know how to use social media and it makes the study process more difficult." Administrators as well acknowledge the need for staff development, which includes the need to develop acceptance of social media, to adapt a social media teaching style, to balance content and tools. In fact, it reveals how challenging social media use for the lecturers is and how much training is needed. It is not enough just to learn technical skills to operate social media tools but it is also necessary to balance teaching and tools. Similarly, Bates (2005) advocates that there is the need to redesign and reorganize teaching in order to successfully and fully apply the new technology, which, according to the author, is often ignored.

The following issue relates to the fusion of formal and informal learning. Student research participants admit that they learn from their peers while sharing the information on social media: "I participate in group discussions, try to find out the answer to the questions together." It seems that more and more students learn in an informal way by navigating social media environments. The idea is also supported by theorists (Siemens 2004) claiming that learning occurs in multiple situations and multiple social media environments and constitutes a bigger part of the learning that a human undergoes during the lifetime. This does not mean that formal education disappears, however, as the teachers acknowledge; their role changes into tutoring, guiding, which is discussed by the researcher Kop (2010), who regards educators as "trusted human filters of information"; it also means keeping the best features of the teacher activator and evaluator of one's own teaching according to Hattie (2008).

The previous issue is followed by the changes in the educational institutions themselves. Administrators mention the redefinition of working places and possibilities of moving lectures into social media environments. It seems that social media is bringing inevitable changes into physical education organization as such. On one hand, as McLuhan (2003) states, the age of technologies and electricity has redefined the nature of work by freeing people from repetitive mechanistic work and opening up possibilities to participate in the society creatively. There are more possibilities to advertise, to attract new students, to reach multiple audiences, and to equip lecturers and students with unprecedented educational possibilities. On the other hand, there are demands for training and time to internalize social media, to note that the genuine human need for communication remains.

Four additional problematic areas mentioned by the research participants emerge: fragmented or disrupted communication, still persisting shortage of technological supply, knowledge commoditization and media addictiveness. According to Bauman (2011) fragmented communication becomes inherent to social media nature. The author discusses liquid modernity which embraces fragmentation and mosaic nature of human existence and communication. The research participants identify the fragmented nature of liquid modernity through the disappearance of direct contact "Direct contact with the students is disappearing in social media." Disruption could be observed even in physical communication as people develop a tendency "to being connected to their iPad or something, being constantly on Facebook, mixing virtuality and reality which gives a feeling they are not communicating."

Another modernity issue is knowledge commoditization "In open databases you can get knowledge free of charge but if you want to get a certified credit you have to pay, so it's relatively free of charge." It resonates with the observation by Mason and Rennie (2008) indicating "knowledge commoditization" and "the student as customer" paying for certification or credits. It seems, however, to be just the regular state of affairs in our heavily consumptive society where everything is consumed, including people (Bauman 2011).

Shortage of technological supply, which teachers have to consider in using social media for education, is another issue: "It's difficult to administer tests because not all the computers work, for example there are 9 computers here and two of them don't work and there are 16 students, so I often give them a paper test." In such cases, teachers have to make choices to rely on traditional paper means, which may seem odd. Having the Internet and computer access is now becoming viewed as one of the fundamental essentials like shelter, water, etc. (Cisco 2011). I notice this in my own perception as I get heavily aggravated if the Internet connection disappears all of a sudden. Being constantly connected to the surplus of information seems to be vital, and I love what one of the research participants said about information: "Information is not rectangular paper pages, it is a metaphysical matter and it exists in social media in abundant amounts; we have to understand this and not to resist but accept and teach our students how to navigate the abundance of information, finding and absorbing the necessary, useful one."

The final problematic area is social media addictiveness—or addiction, as the research participants name it. Even with many users expressing dissatisfaction and non-users expressing fear or suspicion, social media is addictive. It is infectious because of its inherent nature; as McLuhan (2003) reveals, social media is an extension of a human. And human extensions fascinate as they are created with the purpose to extend humans. What is more, as Arnold (2003) points out, technologies are created to answer certain questions but their creation changes the question and automatically the answer, in such a way raising different unexpected issues. As a research participant states, "There are people who say that it's a waste of time to write comments on Facebook, but if you start using social media actively you see that it becomes addictive. It's like irritable smoke when others smoke, but when you start smoking it's already not a smoke, and it turns into a possibility to communicate, relax, stress relief, focusing on other things more productively."

3.2.3 Emerging Positive Effects

In the section of positive effects of social media, again, the results are very similar to the international inductive qualitative research results. Speaking about the use of social media additional to the two identified common types of use for communication and sharing information, there also emerge an educational use for fostering student creativity identified by the teacher research participants and collaboration identified by the student research participants. The educational use for fostering creativity fits into the digitalized Bloom's taxonomy or digital taxonomy addressing the aspects of social media incorporation into education as the highest order of thinking skills, the highest aspect of the teaching/learning objectives. Creativity as an important transversal skill expression and development in the environments of social media is recognized by scholars (Scardamalia 2002; Chai and Fan 2018). Collaboration mentioned by student research participants is also one of the key components and drivers in educational environments acknowledged by scholars. Dabbagh and Kitsantas (2012) provide the percentages of students using social media for coursework-related collaboration (30.7% of students use it to access wikis, 49.4% for SNS, 33.4% for video-sharing, 37.6% for blogs, 40.2% for micro-blogs, and 30.5% for social bookmarking). The data shows that university students are using social media actively and collaborating in the frame of their coursework.

Speaking about previously mentioned modes of social media use—communication and sharing information—research participants report that they communicate with their friends, colleagues, and organizations for personal and professional interests. Apart from personal contacts, teachers communicate with their students and other organizations for educational purposes, and administrators try to advertise their institution and attract new students. Information sharing appears to be a wide area for the research participants. They share new ideas and information on organizations and studies, lecturers share study materials with their students, students share information on study difficulties and also with friends or acquaintances, and administrators provide training.

The category of perceived advantages represents positive social media influence while incorporating technology into university studies. Kop (2010) provides an exhaustive list of social media merits and ways in which modern information technologies—including social media—may influence education. The list of certain features of education stimulated by social media is fully covered by the advantages mentioned by the research participants:

- Increased communication;
- Sharing information;
- Possibility to reach vast numbers of people almost instantly;
- Information visualization;
- Wide access to information;
- Simulation (various types of projects that enable the use of various sensory channels);
- New forms of creativity;
- Economical nature, saving resources.

Teachers identify increased information exchange, wide access to teaching materials, and the possibility to monitor student work and provide feedback. They stress that social media increases the possibilities to share information: "For me it's a space where I can get information easily, to ask and get answers easily." Another feature mentioned is the possibility to reach audiences despite time and space limits, and social media as transactional for time and space: "There are no time or place limits, students can connect whenever, teachers can connect whenever." They also admit that social media gives a broader view of a person through opinions shared and linguistic and visual expression. Teachers feel that social media makes communication better and is attractive to the users; for example, the possibilities of instant feedback and discussions, the fact that all senses are involved and that it is easy to use. They stress the importance of institution promotion, dissemination of educational activities, institution representation, and creating and promoting an image of an institution, making it attractive to students. Teachers mention the economical feature as well, saying that social media enables them to save paper and travel time and makes education more accessible by being more economical: "It's clear that you save on travel, you can participate in conferences not traveling abroad."

Students also mention information availability and presence indication, which enables better communication. Wide access to information and multiple sources of information could be used and multiple users could be reached, which leads to working together and collaboration in exchanging and creating knowledge. Collaborative new knowledge creation stimulates new collective forms of creativity when creativity becomes not only an individual secluded process but also manifests itself in the process of sharing, working together, collaborating and creating.

Administrators stress the importance of wide access to information. They mention that the use of social media in one institution triggers a chain reaction causing other institutions to adopt or improve social media use: "Lithuania is a small country and our staff work in other universities and colleges, so it was like a chain reaction, our institution raised the level at the same time raising the level of other institutions."

All the mentioned features of social media influence higher education. Participators in educational processes face enhanced communication via social media, enhanced access to information resources, engagement of all the sensory channels, collaboration and creation of new digital content. Education participators perceive these features positively as moving higher education toward improvement, toward new quality and new envisioned dimensions.

3.2.4 Phenomenological Existentials

As Illich (1996) states, we live in a "milieu technique"—the environments heavily equipped with technologies—and technologies are changing our ways of life, inviting us to reshape our living. According to the observation by Merleau-Ponty (2002), "Our existence changes with the appropriation of a fresh instrument"; so is our world changed by social media. In the interviews with all three groups of research participants, the existentials of temporality, spatiality, relationality, and materiality clearly stand out, just as social media is transactional in distance and time and is targeted to reach vast audiences (Moore 1993).

3.2.4.1 Temporality

Heidegger (1962) introduced the two notions of temporality: the "world-time," in which the world appears to our being, and "ordinary-time," which denotes subjective lived time. Time resonates in almost all experiences of the research participants: "It takes time…," "It saves time…." Time is experienced in various ways. First, time is intertwined with space: "Speed, a quick access to everything, quick search, quick to reach people," "There are no time or place limits" or "The search for information becomes quicker; earlier when you needed some information first you checked resources at home, then you would go to the library, and not always could find the information. Now the Internet allows you to find the necessary information very quickly." It could be felt how it would take time to check one's home space looking for resources, then going to the library, and now everything is just at one's fingertips performed with a push of a button. Another experience of time is asynchronous and synchronous modes; there is flexibility between them when the limits expand and there is no need to connect to social media at a rigidly identified synchronous time: "Students can connect whenever, teachers can connect whenever."

Time is experienced as additional work load; a tedious duty perceived by teachers is devoting time for mastering social media in their hectic timetables: "Certainly at the beginning until you get used how to apply social media, it seems difficult and takes time, but you have to invest your time when you are learning." Mason and Rennie's (2008) insight reveals that in many cases staff are supposed to educate themselves, which causes additional tension and is usually perceived as additional workload to the staff. Time is experienced as preparation for the lecturers: "I spend plenty of time in order to prepare the material" or "Certainly, it takes time for me to upload the test." Time is also experienced as stress: "I had a lecture where there were 30 people, it was a great stress for me because I got so many questions and while I am answering one question I see that a list of questions is awaiting and I see that a student is already waiting for 10 min, I was all sweaty." Seeing so many questions awaiting in time might never happen in a face-to-face class as there might not be a time gap for questions, some students might feel shy to pose their questions, some questions might be similar so by answering one the teacher might be answering many at the same time, or the teacher might be more flexible in politely refuting some questions if the lecture time limit expires.

There is also Baumanian (2011) puantilistic time when time gets pressed into one point or a dot, a so-called "time implosion": "Sometimes you spend hours in front of a computer screen looking for information and after that you cannot understand what you were doing at the computer, as if it eats up your time." Sometimes time is experienced as lost time, as a loss or a distraction: "You lose time because when you start looking for what you need, somehow you slip into a social network or start looking at a different topic, there are many distractions."

3.2.4.2   Spatiality

As observed by van Manen (2014), spatiality stands out as being closely related to time as space and time are experienced as intertwined. Reflecting on spatiality, we should admit that human beings are not just simple objects inhabiting space objectively; human consciousness acts in a subjective way and it becomes the medium in which things get spatialized. Heidegger observed that our consciousness perceives space as structured nearness, and he speaks about *Ent-fernung* (de-severance, de-distancing) caused by technology.

Initially and for the most part, de-distancing is a circumspect approaching, a bringing near as supplying, preparing, having at hand... All kinds of increasing speed which we are more or less compelled to go along with today push for overcoming distance. With the "radio", for example, Da-sein is bringing about today de-distancing of the "world" which is unforeseeable in its meaning, by way of expanding and destroying [altering] the surrounding world. (Heidegger 1962, p. 207)

Social media is perceived like a space where the information exists: "For me it's a space where I can get information easily, to ask and get answers easily.". By some research participants, however, as it is called a cloud it is experienced as a cloud, with some distrust that the cloud might disappear: "We shouldn't forget that social media is in the air, in the cloud, and if one day a thunder hits this cloud, we could remain without anything, so I'm reserved, for me the cloud is the cloud, technology is good but safety first." It is perceived as somewhat space, somewhat realm in the air—something ephemeral.

Space is experienced in a Heideggerian manner as de-distancing, as a reach: "Very convenient when you get ill you can connect to your students from a distance." One research participant observes that one can reach people from a distance, that there is no need to cover the distance in usual ways of going or traveling: "Some time ago I had to go to work to meet the heads of departments to solve some issues; now when there is email, you don't need to go in order to reply to the questions or send a document." It is a space where daily life activities take place, and it is becoming a living space: "I try to do everything on the Internet, fill in documents, buy things, I imagine if there was a possibility to fit on clothes in the Internet, somehow get a version of an avatar like me, I would use it." And it automatically challenges our physical space or the space perceived as physical: "If a person becomes mobile while learning and can learn not depending on place and time, then there is a question if physical institutions will remain," so the virtual space of social media— the cloud—is perceived as a challenge to our lived physical spaces. Space is also perceived as a place of connection: "I like writing messages on Facebook because you can see immediately who is connected and if he/she has seen the message, then communication seems to be livelier." Though imaginary or not, the virtual realm helps people to feel connected. In fact, our existence becomes divided into two realms: physical reality and virtuality, sometimes intertwining with each other, sometimes challenging each other.

3.2.4.3 Relationality

While using social media relationality is experienced in a twofold way: relations between social media users and human–technology relations. As Evans (2010) observes, people interacting by the means of social media do not simply enter into interaction with other social media users but they also interact with the technology itself. Concerning the interaction among social media users, a process of reduction takes place; as observed by Heidegger (1977), humans as beings with deep essence get transformed into mere functionaries. Reduction of relations is experienced by the research participants: "Communication on social media doesn't feel natural for me because people create their profiles, certain images, and when you communicate, you communicate with created images, it's a feeling that images are communicating between themselves."

What is more, the virtual realm pushes our physical realm into shrinking: "More and more often I meet people who are constantly connected to Facebook, and even when you are with them they still are busy on their phones and communication is disrupted"; as research participants name it, communication disruption or reduction take place. In fact, as Evans (2010) observes, communication undergoes crucial changes; similarly, Heidegger (1977) gives an example of changes in farming when technology brings the need to improve nature instead of maintain it. There is an anecdote of two lovers meeting on a park bench and continuing writing messages to each other. It reveals the changing landscape of human communication.

It is not as simple as reduction, though. Research participants experience presence, being with another in the realm of social media: "I like writing messages on Facebook because you can see immediately who is connected and if he/she has seen the message, then communication seems to be livelier" or "It's very comfortable we have a group on Facebook where we communicate and discuss with group mates, share study material, help each other." There is real help, sharing, discussions in the realm of social media, and it even gets addictive, as the research participants name it and experienced it as addiction "You come home and first you connect to Facebook and start communicating, watching videos, reading articles, and this distracts your attention and it's difficult to disconnect." There is also the lived experience of certain intimacy: "My students admitted that for them being behind the technology screen allows them to open up, to communicate more sincerely." What is more, the experience of being in the public virtual realm of social media gives the feeling of privacy, becoming invisible, dissolving in the virtual community: "In a way kind of privacy appears, in fact you get into public space, but as everyone uses it you remain unnoticed." Presence is felt through language, through mental processes and the language, the images, sound and, self-expression artefacts invite people to hermeneutically interpret the information and perceive the others: "First of all I pay attention to what language people use in their writing, then what moods they express" or "From the photos, from everything you can understand, you can find out the opinions of other people, you can find out more than communicating in an ordinary way, because the person may not openly tell you what he likes, what music, what creative arts..."

Again, as Arnold (2003) observes, technology has a double Janusian face, and what for one is a chance to enhance intimacy and closeness of contact for another is a breach of privacy. What van Manen (2014) observes is the Momus face of technology, which seeks to reveal what is secret, intimate, what breaches privacy. Some research participants express their reluctance to open up private territories on social media: "Maybe it's my character, I don't like showing my photos, telling about myself, and it is too personal." Meanwhile, others seek to acquire more information, seek relationships: "There are a few teachers on Facebook and I like it, I like when I pass someone at university and I know the person from social media." There comes a paradoxical clash observed by Arnold (2003) between public and private, the process of one's working place entering personal space. Teacher research participants express their wish for personal spaces while student research participants require intense communication: "We have one teacher who you can ask on Facebook doesn't matter at four o'clock in the morning, 'Sorry professor, I don't understand' and he will answer and explain, but others are old-fashioned, they stress the importance of their personal space. Maybe if their attitude changed we could communicate more easily."

Ihde (1990) identifies four types of human–technology relations: embodiment, hermeneutic, alterity and background. The experience of embodiment, for research participants, is lived through hiding behind the computer screen, getting in a way invisible: "My students admitted that for them being behind the technology screen allows them to open up," where there is no tactility involved, simply the language, images, sounds, and artefacts of self-expression. Hermeneutically, the users are invited to orient themselves in the environments of social media following the signs and the icons of the software language. And it is not always easy—if the signs get changed all of a sudden, the orientation is lost, the user needs to interpret new signs stepping out into unknown territories: "If the media doesn't change and you get used to it and then it changes suddenly, it's always a shock." Or sometimes the user might feel overwhelmed by the abundance of signs, feel burdened in hermeneutical orientation in social media territory: "I noticed that during the day I need approximately five rubrics but every time I need to search for them because they are hidden somewhere, the system should be user-friendly, not so complicated, especially for the beginning students."

Often technology seems to be an object in the background taken for granted and noticed only when there is some failure or lack of resources. We notice technology when we have to change our usual routines, otherwise we just take objects for granted: "It often happens to me that I plan to watch movie excerpts for gist and all of a sudden the projector switches off and then you have to improvise to think of other activities to be done," or "It's difficult to administer tests because not all the computers work; for example, there are nine computers here and two of them don't work and there are 16 students, so I often give them a paper test."

Speaking about alterity, Latour (1993) and Verbeek (2008) suggest that technological objects have agency; thus, humans experience moral force while encountering technology. Some research participants experience social media as if it were a living thing or has the power of a personality: "Somehow I became afraid of all the networks, because Google search can see what you are doing, all my moves in the digital space are registered" or "I liked the quote which I found that in 2011 there were billions of questions searched on Google, so then there is a question who people would address these questions to before Google." I found even images saying if Google was a guy, he would look like this".

I would agree with Castells (2007) and Irrgang (2005), however, that technological transfer does not automatically lead to modernization; appropriate cultural transfer is necessary. If technological advancement is applied not for exploitation or power exertion, then there is a hope for social advancement.

3.2.5   The Core Super-Ordinate Themes of the Phenomenon

The core super-ordinate themes that form the nucleus of the phenomenon stand out at the deepest level of the analysis. The two key super-ordinate themes encompass the pedagogical relationship between lecturers and students while using social media and lecturer creativity expression while using social media in university study processes.

3.2.5.1   Pedagogical Relationship

The super-ordinate theme at the center of the phenomenon is the pedagogical relationship. No wonder, if we are speaking about educational use of social media that the subject of the pedagogical relationship should emerge in one way or another. Van Manen (2013) indicates that pedagogical moments are the moments when adults have to do something in relation to a young person, or simply when another person enables the latter to develop and grow naturally. Educational environments in educational institutions are the thriving fields for such pedagogical situations. There are usually institutional attempts to record, describe, prescribe, and regulate the pedagogical relationship in various ways, but still the pedagogical relationship seems to be one of most enigmatic fields where Greek *aletheia* shows and hides again to be captured and lived through for a short moment.

Biesta (2013) asserts that nowadays the "learnification" context is legitimized at the international—not to mention national or institutional—level; "learnification" is so far tightly embedded in the EU documents like those from the Bologna Process and the Leuven convention. The author reveals that learning and learner autonomy— seemingly innocent notions—screen or push into oblivion the idea so well developed in the Utrecht school of pedagogical thought that the pedagogical relationship is an asymmetric relationship wherein the educator takes responsibility of the relationship and acts in a responsible way. On the other hand, responsibility should not turn into totalitarian control, of which the threats of the dominating power are well revealed in the panopticon metaphor by Foucault (1998). The question of normality and regulation is well encoded in Greek mythology—a well-known terror of the Procrustean bed.

In the process of unfolding the pedagogical relationship, we could observe a continuum of attitudes and unfolding situations where on one end we can find control, total schematization, an array of prescriptions that should be followed by all means, and on the other end we would see learner responsibility, which is reminiscent of a phrase from a famous novel by Ilf and Petrov that the main character Ostap Bender likes to quote in numerous tumultuous situations: "The saving of the drowning is the responsibility of the drowning ones themselves" (Ilf and Petrov 2011). And somewhere in between a harmonious action of *aletheia* shows itself to disappear again.

Van Manen (2013) reveals that pedagogical experiences can be positive or negative, but, no matter what, they certainly influence us. Here one of the authors remembers her personal experience with her teacher of music. Being a young girl, she really tried to do her homework. But her family didn't have a piano at home, so she had to go to the hospital where her mother worked as a nurse to practice. She practiced as much as she could; however, her music teacher got worked up so much she started shouting at the girl that her hands were not positioned and hold appropriately, that the girl looked like a cow and sat like a cow. The words would hit the heart of a little girl like nails.

Even now being an adult when the author is writing about her experience she can't help crying. Now being an educationalist the author clearly realizes that it was a really bad pedagogical experience, but maybe it enabled the author to see beyond any instructions during her pedagogical career and let her students grow without putting them into a Procrustean bed. Maybe it evoked her genuine interest in the pedagogical relationship, looking for answers if things could be revealed trough research, through writing and sharing and helping others grow pedagogically if, hopefully, they will stumble on my texts in the relentless flow of information. This presents the latency of pedagogical experience well discussed by van Manen (2013).

Pedagogy is in the routine and reflective, habituated and deliberate, preconscious and conscious practice of teaching, and in the pedagogical relation there is influence, exercised by the adult toward the young people (van Manen 2013). The Utrecht school perception focuses on pedagogy as more concerning the child, but if we think about pedagogical experiences and moments in our lives we could notice that there are pedagogical experiences not even linked to childhood but later in life as well, experienced in an asymmetric way from our university professors, our bosses and sometimes even our spouses. Like Buddhist teachings say: Almost anyone we meet in our lives becomes our teacher.

What is more, if we look at the first part of the definition we could see that it is a preconscious or conscious practice, so in the context of educational institutions we naturally have the pedagogical relationship and pedagogical experiences, and this pedagogical relationship is primarily based on teacher beliefs and institutional regulations. Sometimes institutional regulations and official discourses are so internalized by the teachers that they are perceived as their own perceptions and are applied with the genuine belief that everything is being done for the good of the students.

Despite the above, as the writer William P. Young reveals in his novel *The Shack* in one of the insightful dialogs:

But you gotta admit, rules and principles are simpler than relationships.

It is true that relationships are a whole lot messier than rules, but rules will never give you answers to the deep questions of the heart and they will never love you. (Young, 2006, p. 42)

Relationships are characterized by not having certain and rigid rules. Surely, certain rules exist, but at every moment in a real living pedagogical relationship the rules require fresh redefinition as each pedagogical situation is unique in its own right, according to Jespersen (2001), and requires a unique approach each time. Similarly, Freire (2007) reveals that the "banking" metaphor of education is directed toward dehumanization of both students and teachers. The metaphor includes the approach whereby students are treated as empty bank accounts constantly open to the knowledge deposit by the teacher. The author argues that such an approach stimulates oppressive attitudes in society and suggests using an "authentic" mutual approach to education that admits human incompleteness and education as a constant act developing human completeness and mutual relationships between students and teachers in this process.

If we look at the major category of the pedagogical relationship, we can see that the interviews with the research participants reveal a twofold nature of perception of the pedagogical relationship: The first one is a more controlling, regulative approach, and the other is more open, democratic view. As was mentioned above, asymmetry of the pedagogical relationship requires responsibility on the teachers' side, but is control really responsibility? In a way, some control is involved in responsibility because a responsible person would perform himself/herself, but as well they will try to make sure the students in the educational process face certain situations that allow them to learn or develop and grow.

On the other hand, as Lee and McLoughlin (2010) point out, social media brings new, more democratic forms of studying and, there might be some issues faced by the institutions of formal education and institutions of higher education in adopting the new more democratic approaches instead of the linear controlled usual ways (Table 3.26).

In the theme of regulative approach, there are sub-themes related to control, student questions, and multitasking:

- Exercising control;
- Distrust in non-controlled practice;
- Institutional control strategy;
- Attempts to eliminate student questions;
- Non-acceptance of student multitasking/task-switching.

The sub-theme of exercising control includes the research participant perceptions that control remains inevitable in teaching and that they have the ability to observe students' activities. On the controlling approach side, therefore, the research participants express their controlling role, admitting that: "I control, inevitably remains the process of control in teaching" or "I can see how many attempts there were, how much time they spent on the task." The study participants experience control like seeing what students are doing or have done, how much time they spent on the task, how many attempts they had and what result they got. And the students also know that their activity is registered and seen by the teacher. In addition, there is also distrust in non-controlled practice by doubting if student non-controlled practice on social media can be fruitful. Another sub-theme is related to the embedded institutional control, which is revealed in the university institutional strategy to control studies and almost clearly stating that the teacher's responsibility is to control and the learner should be responsible for his/her own studies. This sounds in tune with the notion

Table 3.26 Pedagogical relationship

Meaning unit	Abstracted unit	Sub-theme 1	Sub-theme 2	Theme
I control, inevitably remains the process of control in teaching	Inevitably remains the process of control in teaching	Inevitable control in teaching	Exercising control	Regulative approach
I can see how many attempts there were, how much time they spent on the task	I can see student attempts and time	Ability to observe student activities		
It is very important for them because they know that I see what tasks they did how much they did and what results they got, how many times they attempted the tasks	They know that I see what tasks they did and how			
If you are doing tasks, for example a lecturer uploads them on Facebook, and you are doing the tasks alone at home, the first thing is that nobody controls you, and, for example pronunciation, you can pronounce as you wish	If nobody controls you, you can do as you wish	No control is not trusted	Distrust in non-controlled practice	
Our university adopted ECTS credit system and a strategy that the number of hours for independent studies should be increased but as well independent study hours should be controlled	University strategy that independent study hours should be controlled	University strategy to control studies	Institutional control strategy	
Teacher responsibility is to prepare course material and upload it in the system and then control the process	Teacher responsibility is to prepare course material and them control	Teacher responsibility to control		
There is a desire to move the responsibility to the learner for his/her own learning	Desire to move responsibility to the learner for learning	Moving learning responsibility to the learner		
The responsibility is attached to the student when he/she is doing a test in Moodle environment, the responsibility is that a student is checking his/her knowledge not just using a dictionary to complete the task	Responsibility for checking one's own knowledge			

(continued)

Table 3.26 (continued)

Meaning unit	Abstracted unit	Sub-theme 1	Sub-theme 2	Theme
I put up materials for students so that they would not write to me and ask, they have all the information with the dates and everything, they can do everything interactively and the system shows to me what they did	I put up the material, everything so that students would not write to me and ask	Information provided to eliminate students questions	Attempts to eliminate student questions	
If I have a lecture in a room where there are computers and if I don't tell them not to turn on the computers, everyone gets into Facebook and if I tell them to disconnect, dissatisfaction is expressed	If I don't tell them not to turn on computers, everyone gets into Facebook and is dissatisfied if asked to disconnect	Not permitting student need for multiple channels	Non-acceptance of student multitasking/task-switching	
They are working on my blog, at the same time working on their Facebook; in Facebook they communicate with their friends, relatives, they are always in a kind of background while learning, and they cannot get rid of their multitasking. If Facebook gets stuck, they look at their phones; focusing on just one question is boring for them, they need other information channels to scatter their attention	While doing tasks students need other channels to scatter their attention	Seeing student need for multiple information channels	Acknowledging students multitasking/task-switching	Open, democratic perception
I show everything to my students, where they can find the material, where they can see their results, I even show my environment, that I can see all their work	I show everything to my students, even my environment	Sharing information about lecturer environment	Cooperative atmosphere	
If we have a smart board we do interactive tasks together during the lectures, it gives some freshness to studying	We do interactive tasks together which gives some freshness	Group work		

(continued)

Table 3.26 (continued)

Meaning unit	Abstracted unit	Sub-theme 1	Sub-theme 2	Theme
I am more a facilitator, I allow them to read, to tell, to send to me, record themselves, to find something additional; there is a wish to show the students that there are learning possibilities in social media, not only communication	I am a facilitator allowing students to find something additional	Allowing students to find information	Facilitating/activating student activities	
When we work in Moodle, students have to communicate via forum, someone tries to upload homework, gets scared that cannot see it, I answer and help	A student gets scared and receives help	Help rendering		
The other day a student is sending me a message, that she is on a bus home and is not sure if there will be the Internet connection during the lecture. Then I answer that she can upload the tasks when she gets home if there is no Internet connection on the bus	A student is asking if she could perform the task later and the lecturer gives permission	Flexible decision making	Flexibility	
There were some students trying to avoid doing tasks, they would connect at the beginning of the lecture and then would ask what to do if they cannot manage to finish the task. They didn't know that I could see what they were doing. When we met in face-to-face classes I told them about this and we laughed together	Students trying to avoid doing tasks not knowing that the lecturer can monitor, the lecturer didn't punish	Lecturer tolerance revealed	Tolerance	
Maybe teachers need help, not make the environment so formal, because in Moodle there is certain formality, lecturer can see how the students are performing each task; maybe more informal use of social media, maybe then we could communicate successfully in a less formal way	We could communicate more successfully in a less formal way	Wish for informal communication	Preference of informal communication	

of perceived personal control in university or college studies researched by Perry and Smart (1997); however, they prove that perceived personal control works well only if positive feedback on study results is received by the students and if teaching is provided—otherwise helplessness is felt and students do not achieve. The understanding of the asymmetric pedagogical relationship developed by the Utrecht school therefore stands out here, reminding us that the asymmetric relationship is based not purely on control but also on teacher responsibility.

Another sub-theme reveals teachers' attempts to eliminate students' questions: "I put up materials for students so that they would not write to me and ask, they have all the information with the dates and everything, they can do everything interactively and the system shows to me what they did." On one hand, it is an absolute necessity to keep people informed as much as possible, but I just recall my personal experience, when a new environment was introduced in my banking system, so the managing team even phoned me to inform me that if I have any questions I could use the free-of-charge line any time for consultations and guidance helping to adapt to the new system. I know that feeling myself when you open the machine screen and stare helplessly though all the information is there but you just need to orient yourself. Surely a banking system has the means and the staff employed to help the customer, while the lecturer seems more inclined to save the time as there are many tasks included in teaching routines: planning, preparation, and improvement of the teaching material, real teaching hours in real time and all the additional duties of the faculty. Students' questions are inevitable, however, and while being in the teaching profession one has to accept it as a natural part of the routine. Here one of the authors recalls a time when her colleague was dissatisfied about parents' impoliteness since they called her in the evening after her working hours just to enquire what could be done to improve their son's achievement. In fact, the faculty was divided in half: There were colleagues who would understand the dissatisfaction while some of the colleagues quoted the story of a priest who could be visited at any hour, could be at four in the morning and would still listen to a hurting soul. Surely it was an extreme example, but in a way teachers have to accept that students' questions are inevitable, otherwise, like one seminar teacher said, "If a teacher could be substituted by computer, then he should."

The next sub-theme relates to non-acceptance of student multitasking/task-switching; some study participants feel that multitasking/task-switching interrupts the study process. Concerning this question, the research is contradictory. Some research claims that students who multitask/task-switch during class time have impaired comprehension of course material and poorer overall course performance (Barak et al. 2006; Hembrooke and Gay 2003; Kraushaar and Novak 2010); however, other research tends to support multitasking as an augmenting factor of technology use in the classroom (Young 2006). Fried (2008) suggests that if the teacher does not integrate technology into the lecture it is better to forbid the technology. So it seems that some research participants who disapprove of multitasking/task-switching with technology have the right to eliminate it: "If I have a lecture in a room where there are computers and if I don't tell them not to turn on the computers, everyone gets into Facebook; and if I tell them to disconnect, dissatisfaction is expressed." Here

I would like to quote a seminar presenter who was presenting on the benefits of a communicative approach in language teaching over the old linear approach of the grammar-translation method and was asked the question of whether he would recommend that only the communicative approach should be used. He wisely answered that the best method is the method which works with a concrete teacher in a concrete classroom. As long as the results are achieved and students enjoy their learning, methods and tools could be applied, changed, and intertwined from the broad existing repertoire; it is simply necessary to know and be able to choose and apply, to juggle with ease the broad repertoire.

The other theme is related to a more open democratic approach, which includes the following sub-themes:

- Acknowledging students' multitasking/task-switching;
- Cooperative atmosphere;
- Facilitating/activating student activities;
- Flexibility;
- Tolerance;
- Preference of informal communication.

All the sub-themes relate well to Siemens' (2004) learning approach, which stresses the overall democratic nature of social media, where learning takes place while connecting specialized information sets, tracing connections among multiple fields, ideas and notions, where teachers may become facilitators and advisors regarding which information and ideas to choose in the abundance of information.

Naturally, there are other research participants who acknowledge student multitasking/task-switching as acceptable, constituting another sub-theme of multitasking/task-switching acknowledgment: "They are working on my blog, at the same time working on their Facebook; in Facebook they communicate with their friends, relatives, they are always in a kind of background while learning, and they cannot get rid of their multitasking. If Facebook gets stuck, they look at their phones; focusing on just one question is boring for them, they need other information channels to scatter their attention." Gardner's (1983) model of multiple intelligence shows that multitasking/task-switching is inherent for people with prevailing kinesthetic intelligence. Here I remember a case from my teaching experience, when I had a student who could not sit still during the whole class. From time to time, he would stand up and take notes or do something while standing. Since it was not hindering anyone's learning, I let the student learn in his own way. At the end of the class, he came to me to thank and told me that I was the only person who would not shout at him and force him to sit down.

The next sub-theme is cooperative atmosphere, which includes the sub-themes of openness to students, in which the study participants reveal their willingness to show all the technology environments to students so that they could feel safe and do tasks together and group work. As one research participant says, the study process acquires "freshness" in this way: "If we have a smart board we do interactive tasks together during the lectures, it gives some freshness to studying." Group activity in the interactive social media is welcomed by the teacher.

The sub-theme of facilitating/activating student activities includes allowing students to find information and providing help when necessary. The research participants consciously experience the move toward facilitation/activation, performing the role of a facilitator/activator and getting students to learn via the medium of social media: "I am more a facilitator, I allow them to read, to tell, to send to me, record themselves, to find something additional; there is a wish to show the students that there are learning possibilities in social media, not only communication." Here it could be seen that facilitation/activation is lived through allowing students to process the information and present their products to the lecturer. Naturally, the facilitator/activator role includes providing help when necessary because a facilitator/activator is not just a passive observer. Whether helping determine just what students do with the information or how to navigate the information resources and create meaning for themselves, the facilitator is always here when a student needs to get oriented: "When we work in Moodle, students have to communicate via forum, someone tries to upload homework, gets scared that I cannot see it, I answer and help."

Closely connected to facilitation/activation is the sub-theme of flexibility. In fact, both education and social media (Siemens 2004) require flexibility and readiness for decision making according to the situation: "The other day a student is sending me a message, that she is on a bus home and is not sure if there will be the Internet connection during the lecture. Then I answer that she can upload the tasks when she gets home if there is no Internet connection on the bus." The research participant's lived experience reveals how technology (social media) merges with the best pedagogical practices—teacher flexibility, in this case—and enables teacher–student cooperation in the study process. Similar to what Heidegger (1977) discusses in his book "The question concerning technology and other essays": how a thing opens a new world that includes new structures to be experienced and new meanings to be constructed. Each new piece of technology reveals new horizons and new possibilities to disclose our inner horizons and inhabit the new world disclosed by the new technology—social media.

Similarly, the sub-theme of lecturer tolerance reveals itself in the social media world. One research participant shares her moment of pedagogical experience: "There were some students trying to avoid doing tasks, they would connect at the beginning of the lecture and then would ask what to do if they cannot manage to finish the task. They didn't know that I could see what they were doing. When we met in face-to-face classes I told them about this and we laughed together." Here the teacher chooses laughter and allows students to develop their responsibility and consciousness without punishing them, just telling them that she can see what they are doing on the tasks. It also reveals that there are multiple situations where we have choices to make, and the more horizons are open to us the more informed choices we make. The social media world is proactive in promising possibilities rather than implying limitations. There is a well-known story about the famous mathematician George Bernard Dantzig, who, while a student, was late for a lecture, so he copied what was written on the blackboard, thinking that it was homework. And in two weeks, he solved the problem and presented it to the professor. He had not known that at the beginning of the previous lecture the problem that had been presented on the blackboard was insolvable.

Finally, the sub-theme of preference of informal communication discloses the need for pedagogical contact well-described by van Manen (2013). The following research participant expresses a feeling or deep primordial perception that something genuine pedagogically is missing if the teacher–student relationship remains formalized: "Maybe teachers need help, not make the environment so formal, because in Moodle there is certain formality, lecturer can see how the students are performing each task; maybe more informal use of social media, maybe then we could communicate successfully in a less formal way." Can teachers expect their students to open up in the environment where students know that they are observed by their teachers? Or is it more like the panopticon of Foucault (1998), which, according to the author, becomes even more limiting in the modern technology world where, due to technologies, our inside is made transparent at the same time, limiting us by this totality of control? Research performed by Murray (1985) in the environment of higher education demonstrates that informality of teacher–student interaction favorably affects student learning and achievement. Van Manen argues that pedagogical contact is essential, speaking about the toxicity of the formalized education environment: "Even well-meaning and competent teachers can become toxic teachers in a world where we become impotent and insensitive to the pedagogy of contact in teaching and learning" (van Manen 2013).

3.2.5.2 Teacher Creativity

New technologies change the nature of human work essentially; what is more, in the information and globalized society the changes embrace personal life—social and cultural dimensions of one's personal world. The restructurization and fragmentation of the postmodern world force the contemporary individual to independently make decisions and construct individual reality, and to create one's own personal, social, and work world. Constant learning and personal development become one of the main parts of human life (Glastra et al. 2004), where creativity and its expression become essential.

Human creativity and the ability of the human race to create various technologies have been analyzed by Gadamer (1999) through the Promethean myth. Gadamer (1999) in his interpretation reveals that while giving fire to humanity Prometheus also gives the ability to acquire *technai*—in other words, cultural skills, the ability to create and have a possibility of self-help. He provides a list of creative activities or human arts, including astronomy, navigation, and medicine. Epimetheus did not provide the human race with any qualities that could ensure its survival. When Prometheus realized that helpless humans were doomed to perish without help, he gave them fire, which symbolizes *entechnos sophia syn pyri* (translation: knowledge of art or craft) (Gadamer 1999). According to Gadamer (1999), the Promethean myth embodies human creativity. Stiegler's (1998) interpretation goes further, showing that fire and tools helped humans to survive and improve their condition, and that humans got wound up with their created technology, turning into cyborgs with organic and inorganic parts, constantly reinventing their prostheses and living their prosthetic existence intertwined with technology, inventing technology and being reinvented by technology.

According to Prensky (2014), contemporary education realities have a tendency to undergo fast changes that require an increasing adaptability from teachers and a creative approach to the technologically wired and changing educational environments. Similarly, the problem is described by McLuhan (2003), who states that the times of mechanistic and linear philosophy are over that the linearity has been changed by the simultaneity and concentricity of the digital age with its infinite intersection planes, where all types of media are constantly interacting with each other. So teachers have to enter this simultaneous world involving themselves in the university study processes, the success of which is determined by personal creative expression, and in which the creativity is realized.

In terms of creativity, it tends to be closely associated with artistic, spiritual activities, but modern scholars extend the concept by emphasizing the practical and professional perspectives. Creative people are treated as vitally important resources (Rickards 1994). Qualities of a creative personality are analyzed by the representatives of the humanistic philosophical school of thought; for example, according to Maslow (1967), courage, freedom, spontaneity, and self-confidence allow a person to work in a creative way and achieve self-realization. The approach to creativity and creativity research has been systemized by Sternberg and Lubart (1999). They distinguish seven paradigms of creativity evaluation ranging from the mystical approach to creativity, addressing psychoanalytic, pragmatic psychometric and cognitive social–personality approaches to modern, interdisciplinary theories. The most relevant creativity research paradigm, which is identified by the authors, is the confluence paradigm. The aim of the theories attributed to this paradigm is to reveal the multiplicity of creativity, to combine a wide range of components and to convey the understanding of the broader impact of the context.

Cropley (2008) introduces three main characteristics of the particular importance of creativity in psychology and education. The first is novelty which can be a product, a process or an idea; in our case, it might be the innovative use of social media in university studies. The second one is efficiency, which means that it works and gives some results that can be aesthetic, spiritual, or tangible. The third one is ethical; this is because the term "creativity" is not used to describe the manifestations of self-interest or destructiveness. Cropley (2008) also discusses such important components as creative products, which can be tangible or not, and creative personality as the cause or the potential for the products to appear, as well as creative personality traits such as openness, flexibility and courage, and the interaction with the social environment, where space for creativity appears. The creative personality traits discussed by Cropley (2008) are revealed as specifically important characteristics of teachers using social media in university studies.

Teachers, using social media as new technologies in university studies, demonstrate such qualities as interest in innovation, self-confidence and the ability to experiment and take risks, which are inherent qualities of creative personalities. Personality characteristics and socio-cultural environment as a source of creativity, as key factors of creativity, are identified in the social–personal approach theories. Amabile (1983) indicated that certain personality traits such as independent thinking, self-confidence, and interest in complex phenomena, aesthetic needs and risks often characterize creative people.

Csikszentmihalyi (1996) applied a systems theory perspective for the study of creativity. The author distinguishes three systems, the interaction of which induces creativity. The first system is a system of individual where creative ideas appear as well as the need to create, to change and improve. Every creative act or a piece of work in a certain sense is considered to be a deviation from the norm. When some of these variations are positively assessed they get rooted and become the new norms. The lecturer works in a creative way applying new technologies, new methods and tools. The second system is social environment, which includes people, who promote and accept lecturers' creative initiatives. In a university environment, they are colleagues and the management. The third is culture. The expectations of the faculty performance dictated by the managers and the reaction of the faculty itself shape the unique culture of the organization.

Amabile et al. (1996), while investigating the influence of organizations on creativity, identify three levels of creativity, which show how creativity is fostered. The level of the social environment of creativity recognition begins with the provision of new ideas, their development and promotion of the exchange of ideas, their succeeding support and implementation. The second level of the social environment includes the manager approach and the promotion of creativity at the institutional level. The third level of the social environment refers to the peer support, collaboration, openness to innovation and change, constructive approach to challenge, and the relationship based on trust and mutual assistance. Jeffrey and Craft (2004) stress that creativity is an essential element of self-expression and satisfaction and is vital in ensuring a motivated and meaningful learning and life including lifelong learning in spite of changing market forces and employment opportunities.

Applying social media in university studies, mastering it themselves and using it for teaching, lecturers get involved in the continuous university study processes, the success of which is sustained by the traits of creative personalities and the environment supportive to creativity. Analyzing the external influence, Table 3.27 below represents the theme that reflects the external factors influencing teacher creativity while using social media in university studies.

The theme of social environment effects represents external factors influencing teacher creativity. Social environment effects are associated with colleague influence, exchange of informative ideas and the implementation, and the impact of the university as an institution. The influence of the colleagues is transmitted by their examples and sharing of ideas. This factor is closely related to the impact of informative ideas because ideas can be acquired directly from colleagues at work, as well as at various seminars, courses, and other training sessions, while reading literature or

Table 3.27 External factors affecting the creativity of teachers using social media in university studies

Meaning unit	Abstracted unit	Sub-theme 1	Sub-theme 2	Theme
An example of a colleague, let's say she inspires you by her sincere amazement	An example of a colleague, she inspires	Colleague example	Colleague influence	Social environment effects
There were teaching staff and every lecturer shared this/her own ideas	Every lecturer shared this/her own ideas	Colleagues sharing ideas		
However, I started using it actively when I started training my other colleagues	I started training my other colleagues	Training colleagues		
During the seminars there are plenty of things presented	Plenty of things presented at seminars	Information acquired at seminars	Exchange of informative ideas	
During the courses we were shown lots of Internet addresses where free-of-charge programs are available and we could use them	During the courses free-of-charge programs presented	Information acquired at courses		
You notice some ideas, read something in books, read what others write, was at the seminars in Gratz, you hear what people say and you need to listen to ideas and perceive them as constructive	You read something in books and you need to listen to ideas and perceive them as constructive	Ideas in literature sources		
If I start remembering how I started using social media, when I started using it, I see some ideas from various technical experiences, creating a program, perceiving the logics of the program, and then there was the intranet…	I see some ideas from various technical experiences, perceiving the logics of the program	Ideas acquired through experience		
Which you do not use later, if you do not start using them actively and systematically	You start using them actively and systematically	Active social media use	Active implementation of ideas	
How I started… there were training seminars at our university, so I joined them, we had to prepare an interactive study module, and my colleagues did not have time, so I had to do everything myself	we had to prepare an interactive study module and my colleagues did not have time, so I had to do everything myself			

(continued)

Table 3.27 (continued)

Meaning unit	Abstracted unit	Sub-theme 1	Sub-theme 2	Theme
I started using... Moodle environment most probably it was my first social media because the university forced to use it	University forced to use social media	University institutional impact	Institutional influence	
University encourages the use of social media environment so we start using it	University encourages the use of social media			
Students use Moodle environment because it is formalized social media and the university propagates it	University propagates social media			
Everything started from a workplace situation, some colleagues did not have a computer at home, some of them did not know how to use text editing tools but as it was a workplace situation they were interested in performing the job	A workplace situation colleagues were interested in performing the job			

observing something: "You notice some ideas, read something in books, read what others write; you need to listen to ideas and perceive them as constructive."

Another source of ideas is the person him/herself, when reflecting and summarizing the experience in working with a variety of programs and equipment. Colleague example encourages and "inspires" the person to start using social media; however, active and systematic use is necessary until, through the process of training others, a perception arrives that the media has been mastered: "I started using it actively when I started training my other colleagues."

Amabile et al. (1996) also identify three main factors while analyzing the specific organizational factors influencing creativity. The authors admit the importance of provision and development of the new ideas as well as promotion of the exchange of ideas, idea support, and implementation. Another important factor for creativity in working life is a favorable team, who are defined as open to innovation and collaboration. The study participants, as well, speak extensively about the sharing of ideas and colleague influence: sharing information, training each other. Finally, Amabile et al. (1996) discuss the influence of an institution, which is expressed through manager-encouraged creativity, setting the institutional performance targets as favorable for creativity.

Some study participants indicate the university's institutional impact on encouraging the use of social media, stating that it "encourages the use of Moodle environment so we start using it." Some study participants, however, perceive the university impact as an enforced matter, as a pressure from the institution: "I started using... because the university forced us to use it." Here we can remember Foucault's (1998) paradigm indicating that educational institutions have ritualized their practice regarding discipline and control and that the control extends over all the institutional levels; that is why some study participants perceive institutional impact as a pressure.

Another interesting observation could be drawn from one study participant indicating that sometimes pressure and stressful situations could lead to the exercising of personal creativity: "How I started... we had to prepare an interactive study module, and my colleagues did not have time, so I had to do everything myself." Stressful situations that encourage people to mobilize their inner resources for acting are identified in psychology textbooks as "fight-or-flight" situations, but they concern more instant physical reactions; however, it is highly probable that creativity is mobilized as well, together with the whole body and mind alert. Teachers are affected by various trainings, seminars, literature, colleague examples, diverse technical experience acquired in work situations, and the encouraging influence of the university as an institution.

When talking about how they personally felt their creativity as they began using social media, the survey participants indicated their personal qualities that helped them start using social media, develop their skills, and apply them creatively in university studies.

Inner personality characteristic is the theme that reflects the inner creative personality factors encouraging the use of social media as a new technology in university studies. The theme of personal characteristics is presented in Table 3.28 above. Such creative skills as the ability to take risks allow teachers to confidently try new techniques without any fear of errors in the testing and attempts: "I can take risks, I'm not afraid, I'm not afraid of losing the material." Another related feature is the ability and willingness to experiment: "You try doing something, you don't know, you keep trying ☺ if you don't know, you ask"; such qualities allow mastering new media through trying as if through playing. Another important feature—the ability to solve problems creatively—allows one to see the ways to apply social media when the other technologies do not work. All these features are mentioned in theoretical approaches to creativity as creative personality characteristics (Maslow 1967; Cropley 2008). The Lithuanian scholar Tidikis (2003) also recognizes similar qualities of a creative personality: ability to risk, flexibility and "childishness"—in other words, the ability to play, to experiment. Other important factors that also characterize creative personalities are deep fascination with media, which means the genuine interest in promoting a deeper understanding of social media, and the enthusiasm to apply it: "And you get inspired, and you try."

Finally, insights about the need for information, about the significant changes in the information communication and creation processes, the perception that changes are inevitable and that it is necessary not only to abandon any resistance to change but also to live with it, all of these personality-related factors influence social media adoption and use. Teachers identify creative personality traits, such as the ability to take risks, flexibility, willingness to experiment, enthusiasm, and openness to innovation and change, as helpful in the process of adopting social media and using it in university studies.

Table 3.28 Teacher creative personality characteristics unfolding while using social media in university study process

Meaning unit	Abstracted unit	Sub-theme 1	Sub-theme 2	Theme
Risk level, I can take risks, I'm not afraid, I'm not afraid of losing the material, I know I can find it, I'm not afraid that it will take some time to connect everything, I will find out how to do anything	I can take risks, I'm not afraid of losing the material	Ability to take risks	Creative abilities	Personal characteristics
It seems difficult, you try doing something, you don't know, you keep trying ☺ if you don't know, you ask	You try doing something, you don't know, you keep trying	Ability to experiment		
In other media fields I'm still experimenting	I'm still experimenting			
When technologies would let me down, then I would try to get around the problems, to get to the Internet. If I can't present my teaching material, I always have it on the net, in my blog, students can open it on their computers, and I don't need to send it separately to everyone	When technologies would let me down, then I would try to get around the problems, to get to the Internet	Ability to solve problems creatively		
We got interested in social media	Interested in social media	Interest		
It first started when I began reading newspaper and magazine blogs and these blogs became more interesting for me than the publications themselves. They are faster more interesting and more active, maybe the information isn't verified on blogs but it is more personalized and more interesting linguistically and appropriate as language learning material for students	Blogs became more interesting for me than the publications themselves, and more interesting linguistically			
And you get inspired, and you try, and you want to share with others...	You get inspired, and you try	Enthusiasm		
Information is not just square sheets of paper, information is a metaphysical thing, it just needs to be understood, it exists here in great quantities and it is important not to oppose to the flow	Information is not just square sheets of paper, information is a metaphysical thing and it is important not to oppose to the flow	Openness to information flow	Openness to changes	

The research reveals two themes determining teacher creativity while using social media in university studies: external factors that promote the expression of creativity of the lecturers while using social media in university studies and the inner creative personality characteristics that allow lecturers to successfully use social media in university studies. The theme of external factors includes colleague influence, sharing ideas and their active implementation, as well as the institutional influence of the university, which is perceived by the research participants as institution encouragement or pressure at times. According to Csikszentmihalyi (1996), from the systems theory perspective on creativity there are three interacting systems: the individual, the social environment and culture. The theme of external factors distinguished in the research unites the manifestations of the social environment and culture systems. The social environment theoretical system is related to the topic of colleague influence distinguished in the research, and culture system is related to the institutional influence of the university. Faculty performance expectations dictated by managers are revealed in the statements of the research participants that the university encourages and propagates the use of social media.

Speaking about the external factors, teachers state that such factors as colleague examples and new ideas that teachers want to try are stimulated by the internal factors—personality characteristics such as willingness to take risks, willingness to experiment, openness to innovation and a desire to creatively solve problems. All of these features and their expression are manifested in the process of teachers using social media in university studies. These themes are intertwined together and form a whole system of factors influencing the process of teacher mastery of the media and the use of it in university studies. Similarly, personal effects and social structure act as the factors integrally related to each other, as Csikszentmihalyi (1996) states in the systems theory perspective on creativity. Teacher creativity is encouraged by their colleagues, and idea sharing and the faculty performance expectations dictated by the university as an institution encourage the use of social media. We can see that broad creativity understanding delineated by the representatives of the humanistic philosophical approach is manifested in teacher activities through openness to innovation (social media) and the challenges (mastery and application of social media in university studies), continuous learning, and the ability to live and work in the context of permanently changing circumstances. It also highlights the importance of creative personality characteristics, the qualities that promote the adoption of new information, and the use of it in different ways.

Conclusions

Views on Social Media in University Studies from the Literature Overview

Social media embraces numerous applications such as wikis, blogging, social networking, and podcasting. The "social media" definition is still changing as new forms of social media appear and the existing forms are constantly modified. Scholars provide numerous definitions for the concept of social media that are related by the main idea that social media is "a group of Internet-based applications that build on the idesological and technological foundations of Web 2.0, which allows the creation and exchange of user-generated content" (Kaplan and Haenlein 2010). The common thought relating definitions of social media is the blend of information technologies and social interaction leading to co-creation of content and knowledge. User-generated content becomes the main feature of social media; what is more, the content could be changed, redefined, improved, and modified by multiple users. Technology allows users to connect in the process of content creation and, using multiple channels, constantly modify and change it.

In university studies, social media could be best presented by the pyramid in Bloom's revised digital taxonomy, which provides various social media tools grouped according to the taxonomical levels and provides a hint of how some of the social media tools, which match the taxonomical level, could be integrated into educational activities.

It also should be noted that LMSs like Moodle have social media features, although scholars note that LMSs give limited opportunities for online sharing and collaboration as student interaction activities are restricted to one class or one semester as compared with a constant opportunity of sharing many-to-many. Although LMSs have social media features, there are also institutional security and privacy requirements that do not allow sharing beyond the limits of an institution; nevertheless, technically there are possibilities of merging LMS and social media tools together,

G. Valunaite Oleskeviciene and J. Sliogeriene, *Social Media Use in University Studies*, Numanities - Arts and Humanities in Progress 13, https://doi.org/10.1007/978-3-030-37727-4

thus blending the technologies together and making them fall into the same group of social media tools.

Scholars also foresee an extensive growth of social media technology use in university studies due to the numerous factors such as rapid technological change—which naturally pervades all spheres of life including education—increase in student IT skills as the new generation acquires the skills naturally, and growth of demand for university studies to satisfy a growing demand for mass higher education.

At the same time, social media incites changes in university studies. Scholars identify interactivity of learning as an educational technique that requires change in education and list the areas of change: from linear to hypermedia learning; from instruction to construction and discovery; from teacher-centered to learner-centered education; from absorbing material to learning how to navigate and how to learn; from one-size-fits-all to learner-tailored learning; and from the teacher as transmitter to the teacher as facilitator or even more an activator (Hattie 2008). In fact, social media use has a tendency to shift an educator's focus from a teacher to a facilitator and activator of learning, which does not mean that a teacher ceases to teach—it just means that pedagogical focus shifts toward a more democratic one.

Social media has already affected and continues to influence university studies in both global and local contexts so the use of online technologies such as social media is turning out to become a significant challenge for academic staff. There is a certain pressure from the students on the teachers to use new technologies even if explicit institutional policies are lacking, and there is also some growing enthusiasm among academic staff as well. The scientific literature envisions such characteristics of social media use in university studies which are promising increased learning and student engagement, as well as collaboration.

The problem of interplay between new technologies and pedagogies remains the issue to be researched. Researchers admit that there should be developed certain theoretical and research insights on the role of technology in teaching and learning as there is a qualitative difference between "teaching online" and "putting a course online." What is more, learning in social media context is based more on collective efforts of exploration and innovation, and the characteristic of formal education relying on individual instruction is less preferred. These observations resonate with the latest approach of connectivism which stresses the ability to actively access information and augment it rather than rely on the methods focusing on the passive retention of information which were used as traditional ways of teaching and learning in the environments of formal higher education. Many scholars point out that universities have the potential to use social media for collective knowledge creation. However, there are some critical attitudes stating that students while using social media "are evolving from cultivators of personal knowledge into hunters and gatherers in the electronic data forest." Nevertheless, many researchers observe that such critical attitudes are rarely based on extensive research.

It is alleged that the previous lessons learned in the process of applying technology in education should be kept in mind while implementing new emerging technologies; however, the earlier lessons with the technology application are often ignored. For instance, the need to redesign and reorganize teaching in order to achieve full and successful application of the new technology is often ignored. Additionally, it should be kept in mind that technologies do not simply roll in effortlessly, and there always need to be attempts to address certain groups of people making sure they get access to the technology. New cohorts of students enter university studies, and the essential observation is that they are the students who are more willing to multitask, and are used to digital juggling of their activities and increased autonomy of social activity, which equips them with the possibilities of being able to choose what they do, when, where, and how. However, it is identified that traditional top-down institutions (universities) are poorly accommodated to meaningfully engage their students, as scholars observe, "even the best-intentioned universities are able only to offer their students an artificially regulated and constrained engagement with social media." There is a certain incongruity between hierarchically structured way of communication and learning offered by universities as institutions and the linear democratic nature of social media.

A phenomenological perspective on technology reveals that on the one hand technology brings more comfort into human life; on the other hand, it challenges human existence by defragmenting and destructing the existence itself, by becoming a technological plague. From the techno-genetic phenomenology point of view, humans became more and more reliant on artificial means. They apply tools and technology as artificial body parts or so-called prosthesis and are doomed to accept their existence as of prosthetic beings—cyborgs, who are intertwined with technology, create technology, and are affected or created by technology. The development of cognitive technologies and the emerging new media pose a warning against capturing people's attention simply toward commerce of technical industry.

Network approach identifies that technology and social conditions are intertwined recognizing that both factors of technology and social condition intertwine cause and effect at the same time. It presupposes a further theoretical move not separating humans and technology but viewing them in binary connection. In a similar way echoing the fundamental observation by Heidegger that technology does not change the world but it enframes the world in a certain manner, there is a further observation that humans apprehend the world through technological frame, and by seemingly answering a need or a question technology changes the question and the answer at the same time; for instance, social networks not only function as means of enhanced communication, but also change the understanding of community and communication itself inside the community. Technology acts not exceptionally as a tool but on higher metaphysical level. In educational environments (university studies), ICT, including social media, used in the classrooms totally affects and reshapes teaching relations with students and ways of perceiving and interpreting the world. Technology used in educational environments immediately reforms and deforms our living world at the same time inviting to conform to the new horizons of our living world.

Overview of the Research

The inductive qualitative research with a phenomenological approach used while researching social media use in university studies enabled us to travel deeper into the layers of the phenomenon, like in Kvale's traveling metaphor. The first layer opened up through international inductive qualitative research as a starting point of the research and then through inductive qualitative research with a phenomenological approach at home institutions brought out the key contradictions and positive effects of social media use in university studies. The emerging contradictions could be characterized by the Janusian metaphor, speaking about Janus-faced contrasting binaries. The first of these binaries is the abundance of information on social media, which evokes the necessity of media "literacies": attention, participation, collaboration, network awareness, and critical consumption.

Another contradiction is related to the democratic scattered nature of social media and the institutional nature to control and categorize processes. Although educational institutions have ritualized their practice regarding discipline and control, and it seems to persist, the application of social media acts as a driving force toward democratization of educational processes. Moreover, social media brings about another contradictory area—fragmentation of identity. On the one hand, identity remains the object to be hidden, not expressed; on the other hand, language used, photographs, and likes and dislikes reveal the part of identity that might not be explicitly expressed. This is followed by the privacy issue well defined by the myth of Momus, who became the first to express the desire to access what was hidden in the human heart by means of a technology of surveillance. The issues of changing teacher and student roles are related to the changes in university study processes: teachers adapting the facilitator role more often and students becoming more autonomous learners, moving toward fusion of formal and informal learning and bringing inevitable changes into physical education organization as such, redefining working places and moving lectures into the space of social media, and bringing on knowledge commoditization. The latter is a more general modernity issue related to consumptive society—social media only makes it more visible as it enables faster and easier access to the processes of treating knowledge as commodity. One more issue mentioned is shortage of technological supply and Internet access becoming like one of the fundamental essentials. Finally, there is the issue of social media addictiveness or media being a human extension that wraps humans into its "cyborgian" existence.

Positive effects of social media use in university studies include, firstly, how social media is used for communication, sharing information, fostering creativity, and increased collaboration. Social media is used to communicate broadly for personal, professional, and teaching/learning purposes, including sharing and exchange of information through multiple social media channels. The educational use for fostering creativity fits into the digitalized Bloom's taxonomy, or digital taxonomy, addressing the aspects of social media incorporation into education as the highest order of thinking skills, the highest aspect of the teaching/learning objectives.

The super-ordinate theme of perceived advantages of social media use in university studies covers the following list:

- Increased communication;
- Sharing information;
- Possibility to reach vast numbers of people almost instantly;
- Information visualization;
- Wide access to information;
- Simulation (various types of projects that enable the use of various sensory channels);
- New forms of creativity;
- Economical nature and saving resources.

The research participants disclose the abovementioned advantages in their interviews. They reckon that wide audiences could be reached immediately and that there are no space or time limits while accessing and exchanging information. They also mention that social media helps to save resources such as paper and travel expenses. In addition, information visualization becomes essential, and while applying simulations various sensory channels are employed, which in its own right stimulates new forms of creativity.

Another layer opened up through deep phenomenological study at home institutions is bringing out phenomenological existentials: temporality, spatiality, and relationality. Concerning temporality, there is subjective lived time, which is experienced by the research participants in various ways. First, it is perceived as being interconnected and intertwined with space. Time is experienced as additional workload and also as stress. Time is also perceived as Baumanian puantilistic time when time gets pressed into one point or a dot. Finally, time is perceived as lost time, as a loss or a distraction.

Speaking about spatiality, space is experienced as de-distancing, which implies bringing near or having at hand. Social media is also perceived like a space where the information exists; however, by some research participants it is called a cloud and it is experienced as a cloud, with some distrust that the cloud might disappear. Social media is experienced as a space where daily life activities take place; it is becoming a living space. As a virtual space of social media, the cloud is perceived as a challenge to our lived physical spaces. Space is also perceived as a place of connection. In fact, our existence becomes divided into two realms: physical reality and virtuality, sometimes intertwining with each other and at times challenging each other.

Relationality could be revealed in a twofold way: relations between social media users and human–technology relations. What concerns the interaction among social media users is a process of reduction takes place which means that humans as beings with deep essence get transformed into mere functionaries. Reduction of relations is experienced by the research participants: "Communication on social media doesn't fell natural for me because people create their profiles, certain images and when you communicate, you communicate with created images, it's a feeling that images are communicating between themselves." In addition, virtual relations squeeze out real physical relations as even being physically together people remain busy with their communication on social media. However, not only reduction is experienced. Research participants experience presence, being with another in the realm of social media. There is also lived experience of certain intimacy as being behind the technology screen allows people to open up and being in the virtual realm of social media gives the feeling of privacy becoming invisible. Technology has a double Janusian face or is Janus-faced, so it is perceived by some research participants as a possibility to enhance intimacy while for others it means a breach of privacy. Also, the process of one's working place entering personal space takes place.

Taking into account the four types of human–technology relations identified: embodiment, hermeneutic, alterity, and background, we see that all of them could be found in the discussions of the research participants. First, the embodiment is lived through by hiding behind the computer screen; then, the users of social media employ hermeneutics by following the signs or the icons of social media. Often technology seems to be an object in the background taken for granted and noticed only when there is some failure or lack of resources. What concerns alterity is some research participants experience social media as if a living thing or having a power of a personality.

And finally at the deepest layer, we find the pedagogical relationship and teacher creativity. The university study process of teaching/learning still has the core element of the pedagogical relationship—the relationship between teacher and student—which persists in technologically equipped environments. The asymmetry of pedagogical relationship involves certain responsibility on the teachers' side. In a way, some control is a necessary part of responsibility because a responsible teacher will try to make sure the students in the educational process face certain situations which allow them to learn. On the other hand, it is perceived that social media brings new more democratic approaches to studying and institutions of higher education, universities in our research, experience the necessity to adopt the new more democratic forms instead of common linear controlled ways. In the research, the continuum of pedagogical relationship opens up featuring at one end regulative approach based on belief in linear control of university study processes and at the other end embracing a more open democratic approach. The regulative approach according to the research findings includes such characteristics:

- Exercising control;
- Distrust in non-controlled practice;
- Institutional control strategy;
- Attempts to eliminate student questions;
- Non-acceptance of student multitasking/task-switching.

Exercising control includes the type of thinking that control remains inevitable in teaching, it also includes the belief that observation and minute registering of students' activities enhance learning, and there is also certain distrust in non-controlled practice by doubting if student non-controlled practice on social media can be fruitful. The embedded institutional control is incorporated in university institutional strategy to control studies by explicitly stating that teacher responsibility is to control.

Also, teachers' attempts to eliminate students' questions show the inclination to control, to create a purely clinical controlled environment where students' questions are treated as undesirable or as a failure to provide all possible provisions for learning. In addition, student multitasking is viewed as more an impediment or impairment of learning.

A more open democratic approach in the pedagogical relation includes the following features:

- Acknowledging students multitasking/task-switching;
- Cooperative atmosphere;
- Facilitating/activating student activities;
- Flexibility;
- Tolerance;
- Preference for informal communication.

All these features could be related to the connectivism learning approach, which is based on the idea that the democratic nature of social media provides more open possibilities for learning by connecting multiple fields of information where teachers become more facilitators or advisors on how to navigate the ocean of information and ideas. In such environments, student multitasking, working with multiple sources and ideas, seems natural. Equally natural seems the overall cooperative atmosphere where the teacher acts more as facilitator and demonstrates flexibility and readiness for decision making according to the situation. The social media world is proactive in promising possibilities rather than implying limitations. Teacher tolerance and informal communication find the way as the best practices of pedagogical contact, encouraging students to open up in the process of learning. It is like a Heideggerian thing that opens a new world, which includes new structures to be experienced and new meanings to be constructed. Each new piece of technology, social media as well, reveals new horizons and new possibilities to disclose our inner horizons and inhabit the new world disclosed by the new technology.

Another observation is that human creativity persists even in a Stieglerian cyborgian existence of human existence. Human existence might be shaped by technologies, but the source of genuine creativity is present in humans themselves; so whatever the environment and whatever the situation, human creativity will spring out looking for the ways to inhabit technologized environments. Two themes related to the factors affecting teacher creativity are distinguished: external factors that foster the expression of creativity of the teachers while using social media in university studies and the inner creative personality characteristics that allow teachers to successfully use social media in university studies. The external factors include colleague influence, sharing ideas, and their active implementation as well as of the institutional influence of the university which is perceived by the research participants as institution encouragement or pressure at times. Colleague examples and new ideas that teachers want to try are stimulated by the internal factors—personality characteristics such as willingness to take risks, willingness to experiment, openness to innovation, and a desire to creatively solve problems. In fact, these themes are intertwined together and form a whole system of factors influencing the process of teacher mastery of the media and the use of it in university studies.

Discussion

Both the literature review and the findings of the research indicate that the phenomenon of social media use in university studies is a complex phenomenon with multiple layers. First, the rapid development of multiple Web 2.0 technologies, including social media, challenges the Cartesian paradigm—the linear deterministic cause-and-effect way of thinking (McLuhan 2003); however, it is not so evidently easy to change or switch from the deterministic approach as it is still widely applied even in creating technologies themselves. It is evident from the literature overview that there remains the prevailing tendency to measure the benefits and drawbacks of social media, and to classify and provide exact measures. The first layer of the phenomenon revealed in the research, however, shows that though there are certain benefits the contradictions identified acquire a paradoxical character that is not in tune with the linear Cartesian paradigm; for example, social media opens up access and continuous creation of the abundance of information, but research participants sometimes feel that it is almost impossible to grapple with the information flow.

Then, there appear new theories on media "literacies" and demands to equip media users with them. It reminds us of a story read in a newspaper about a consumer in the USA suing the ice-cream producers for causing stress and disappointment because he was devastated when he saw how many flavors of ice cream are constantly being produced and calculated that even if he tasted a reasonable number of ice-cream flavors each day his life still would not be long enough to taste everything. All the Janus-faced contradictions of paradoxical nature hardly fit into a linear good bad, benefit–damage, and cause–effect paradigm; thus, it makes it relevant to overview the philosophical research approach in the first place. Technology operates at a metaphysical level (Arnold 2003) as it enframes the world (notion used by Heidegger) in such a way that the question of technological effect is changed alongside with the answer. The best example could be the Internet, which changes the communication between community members but at the same time it changes the understanding of the community itself, or the Heideggerian example about de-distancing caused by technology, which also destroys closeness. In this metaphysical light, a phenomenological approach seems relevant, bearing in mind Gadamer's (1999) statement that we cannot reach the finite knowledge, as it is interrelated with the context, but we can reach certain iterations of knowledge.

The above, however, does not make linear perceptions of positive effects and problematic issues caused by social media application invalid. It just allows us to perceive linear perceptions as one of the multiple layers of the phenomenon. Returning to the overwhelming outpour of information through social media, Rheingold (2010) identifies that not all the information is appropriate all of the time; in many cases, there exists "rubbish" information, as the research participants call it. The author suggests the necessity for media "literacies," which include: attention, participation, collaboration, and network awareness and critical consumption, which may help the learner to effectively use social media. In addition, Siemens (2004) focuses on the importance of connecting specialized information sets and the ability to navigate the surplus of knowledge. Similarly, Owen et al. (2006) argue that the possibility to access diverse resources, including resources of high quality, in social media environments may encourage people to develop their critical thinking and enrich thinking and knowledge-managing processes. The research participants speak about the difficulty of managing the continuous flow of information, fragmentation of information, and the scattered character and constant need to sort out, and find and choose the necessary information. They also mention the necessity of developing critical thinking skills. Even with the theoretical media "literacy" answers, though, the paradoxical nature of the contradictions is lived through by the research participants.

Returning to phenomenological approach which admits the defragmentation of the technologically enframed human world, the Janusian myth is well applied by Arnold (2003) to reveal simultaneous binaries provoked by technology, the ever-changing situation when the answer changes the question in its own turn. Such binaries include fragmentation of identity when, according to one of the research participants, "the created images communicate among themselves," while others see visuals, sounds, and language shared revealing a lot about the identity and fragmentation of privacy. When some research participants feel absolutely private hiding behind the screen of social media, others want to access what is hidden like in the Momus myth presented by van Manen (2010). Additionally, phenomenological existentials reveal the change of experiencing the space, so-called de-distancing, closely intertwined with time, which in some cases becomes puantilistic, according to Bauman (2011), squeezed into a dot, while in other cases related to the asynchronous mode where time becomes scattered in the space. Also, fragmented communication is reduced to a picture as seemingly communicating with vast communities in social media people at times do not engage in face-to-face communication; they prefer to remain in the environment of social media, as one of the research participants observes.

The next key matter to be discussed is communication and dialog—the core of the pedagogical relation, according to Biesta (2013). Freire (2007) also values communication and dialog as the key components fueling transformational education and learning. Meanwhile, Weller (2007) admits that there are two different approaches to learning enhanced by information technologies: One attaches more emphasis on content, and the other emphasizes communication. Student research participants put an immense importance on their communication; they especially stress their need for more informal communication with their teachers through various and multiple channels of social media. In fact, the preference for informal democratic contact in education is expressed—they favor a pedagogical relationship based on informal teacher–student communication, cooperative atmosphere, tolerance, and acknowledging student multitasking. Van Manen (2013) argues that even competent teachers may become "toxic" in the environments insensitive to communicative and dialogical pedagogical relationships.

Here, we arrive at the sometimes-fatalistic phenomenological perception of a technological frame or cyborgian existence in a defragmented world. Adams (2010), for example, argues that educational technologies cannot be viewed as neutral artefacts by analyzing PowerPoint use in the classroom and how the teaching experience is limited by the number of slides as well as how the sequence of slides defines the teaching narration and a common answer to a question could be: "I'll answer the question when I get to a particular slide." Furthermore, in most cases the answer is never given and the question has been forgotten by the time a particular slide has been reached. Despite these tendencies, however, even if it seems so easy to follow the framework defined by technology, humans have choices. They have creativity to decide how they are going to react to a question: to break out of the enframed reality and change the narration or simply follow the frame. Even if social media frames educational study environments, teachers still have choices as to how to use social media.

Closely related to the above is the super-ordinate theme of teacher creativity while mastering and applying social media in university studies. On the one hand, philosophers argue that the human world becomes enframed by technologies; on the other hand, the well-known Promethean myth carries the message that the only exceptional feature humans have is creativity—symbolically, the fire given to humans by Prometheus. Such choices humans face in almost every meaningful activity, be it technology enframed or not; that is why thinking outside the box or creativity is so greatly valued. The super-ordinate theme of teacher creativity while mastering and using social media for teaching reveals such qualities of a creative personality that promote the adoption of new information and the use of it in different ways.

Finally, blending multiple approaches might equip us with better understanding and provide us with possibilities of better choices.

Recommendations

The starting point of the recommendations could be the thought expressed by Castells (2007) that people have the possibility to choose how to use technologies instead of limiting themselves to technology-delineated frames or limits.

> The promise of the Information Age is the unleashing of unprecedented productive capacity by the power of the mind. I think, therefore I produce. In so doing, we will have the leisure to experiment with spirituality, and the opportunity of reconciliation with nature, without sacrificing the material wellbeing of our children. The dream of the Enlightenment, that reason and science would solve the problems of humankind, is within reach. Yet there is an extraordinary gap between our technological overdevelopment and our social underdevelopment. Our economy, society, and culture are built on interests, values, institutions, systems of representation that, by and large, limit collective creativity, confiscate the harvest of information technology, and deviate our energy into self-destructive confrontation. This state of affairs must not be. There is no eternal evil in human nature. There is nothing that cannot be changed by conscious, purposive social action, provided with information, and supported by

legitimacy. If people are informed, active, and communicate throughout the world; if business assumes its social responsibility; if the media become the messengers, rather than the message; if political actors react against cynicism, and restore belief in democracy; if culture is reconstructed from experience; if humankind feels the solidarity of the species throughout the globe; if we assert intergenerational solidarity by living in harmony with nature; if we depart for the exploration for our inner self, having made peace among ourselves. If all this is made possible by our informed, conscious, shared decision, while there is still time, maybe then, we may, at last, be able to live and let live, love and be loved. (Castells 2007, p. 396)

There are various attitudes toward social media use in university studies, ranging from extremely enthusiastic beliefs to rather reserved ones; however, it is evident that social media has already entered the environments of university studies, and even newer technologies will continue to pour into our lives, including our education. Educators inevitably face social media application in teaching/learning processes, and they do their best to learn and get used to new applications and to changing work conditions.

The first recommendation that has resulted from the research is that application of social media makes it imperative that teachers use all their creativity in acquiring new skills of using social media in the processes of university studies; however, in many cases staff are supposed to educate themselves and equip themselves with the newest skills, which is perceived as additional workload and causes additional stress. It would be advisable to provide constant staff support and allocate time for practicing and mastering new skills. Also, the best practices of colleagues informally helping each other with the newest skills should be supported and fostered.

The second recommendation based on the findings of the research is related to the fact that considering a pedagogical relationship based on less formal communication with students and adapting more encompassing, engaging, and tolerant attitudes toward students' learning activities might be a challenge sometimes. The controlled environments of formal education give an onset of control and belief in the rigidity of control—such is the essence of bureaucratic systems. The damaging effect of overall control has been analyzed by scholars, and it is recommended to be avoided. Universities provide formal education, and naturally the study processes are regulated and controlled. Research participants—both teachers and students—express their wish for less formalized communication between educators and learners. The importance of a sensitive pedagogical relationship that allows students to learn and develop, taking into account their needs, cannot be denied. Alongside with equipping staff with the newest skills of social media application, there should be allocated resources and time for equipping staff with the newest philosophical–pedagogical insights on application of the newest Web 2.0 technologies and their applications in university studies.

The third recommendation based on this research encourages educators of all levels, including administrators and lecturers, to consider equipping students with social media "literacies": attention, participation, collaboration, network awareness, and critical consumption, and especially focusing on developing students' critical consumption. Student research participants themselves mention that they feel the

need for developing critical thinking in order to be able to cope with constant information flow. Also, teachers could be viewed as "trusted filters of information" or
"knowledgeable others." Fostering critical thinking as one of the higher mental skills
in Bloom's taxonomy should find a place in the university staff teaching repertoire.

Implications for Future Research

The phenomenological approach of the research allowed us to get into layers of the
phenomenon of social media application in university studies; however, as Gadamer
(1999) argues, knowledge is closely intertwined with the context, and finite knowledge hardly exists. In this case, we started with inductive qualitative research across
five European universities and one University of the Third Age, and then we got
deeper into the phenomenon, applying a phenomenological approach and carrying out extensive interviews at home institutions. Another area of further research
could include extensive interviews in some of the other universities where the initial
research has been carried out and cross-comparison of the results.

Also, there is an ongoing theoretical discussion on social media use in university studies. Selwyn (2012) observes that the ongoing debate is still not based on
substantial research and is of a more speculative nature. There are still many unanswered questions on social media use in university studies, and what it makes even
more complicated is the Janusian nature of technologies; human use of technology
creates binaries of opposites where the question is changed along with the answer.
Selwyn (2012) asserts that the wider context of social media use in university studies remains contradictory. First, different conditions are found in access to Internet
and social media tools and digital divide remains great depending on socioeconomic
status (Jones and Fox 2009). Also, democratic activity of social media appears to
be questionable and social media environments are not more socially integrated
than the offline ones (Mayer and Puller 2008). What is more, not all social media
activities are related to educational contexts (Selwyn 2012). Another concern is that
optimistic expectations about social media-enhanced collective creativity seem to be
far-fetched. The majority of users of social media applications prefer passive use of
knowledge, and user-creative activities are mostly limited by profile creation. Selwyn
(2012) identifies two major issues:

1. The discussion on the nature of the institutions of formal higher education (universities in our case) including debates about the nature of institutionalized
   education;
2. Integration of social media into educational environments.

Concerning the nature of institutionalized university education, social media being
a disruptive technology poses an additional challenge in the debate on institutionalized education, where the opinions, philosophical approaches, or political decisions
of what is worthwhile education or worthwhile learning prevail (Standish 2008). I

agree with the claim by Selwyn (2012) that the nature of institutionalized university education opens a wide research field.

Another broad research area is integration of social media into the environments of university studies, which raises more questions than provides answers. There are questions like how collaborative student work could be assessed, quality control mentioned by the administrator research participants, how best to design the curricula inclusive of social media, and how best to support staff and students providing meaningful educational use of social media in university studies. Crook (2008) raises similar questions when he speaks about the need for a sound basis and governance of learning in technologically driven environments.

Final Conclusion

Finally, it is a practical idea to present the conclusions in a structural and condensed way, showing their relationships to the research objectives. Figure A.1 shows the

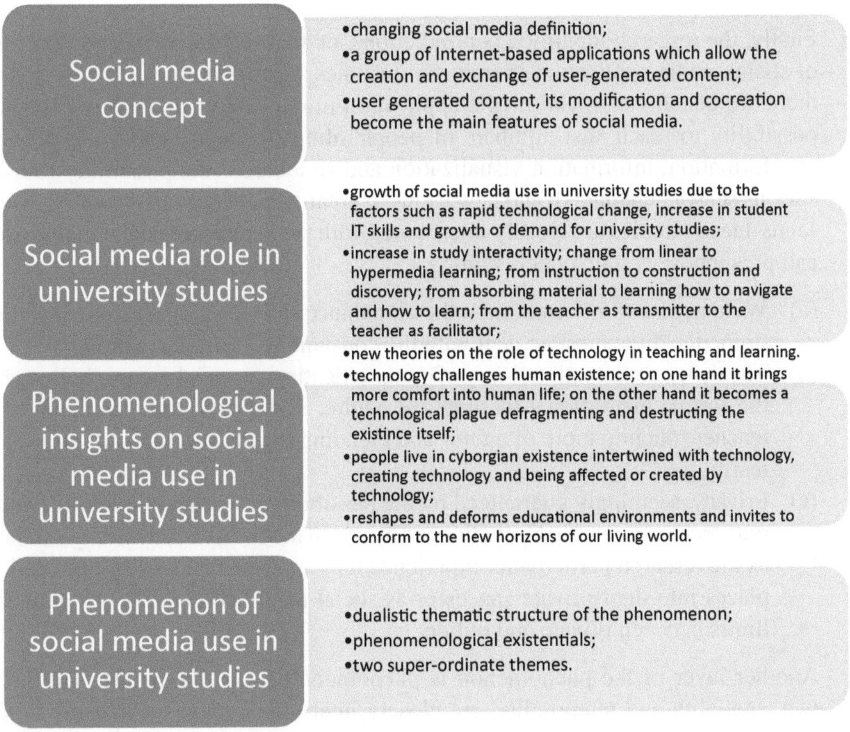

Fig. A.1 Visual presentation of conclusions

visual presentation of the conclusions. Based on the research results, the following key observations can be made:

1. Due to the power of dualistic perception, the first layer of the phenomenon represents positive effects and contradictions of social media use in university studies. The contradictions, however, acquire the binary Janus-faced characteristic of uncertainty and paradox when social media use in university studies demonstrates unintended consequences.
2. Social media use in university studies causes changes in experiencing time, space, and relations, which acquire new (both positive and negative) meaning.
3. At the heart of the phenomenon, there are human creativity and the pedagogical relationship, which essentially belong to the human living world, not to the technology world.

Social media is becoming more and more pervasive in the environments of higher education; it is becoming embedded in university studies. The research question raised at the beginning of the research was: "What is the phenomenon of social media use in university studies?" As the final touch, therefore, we would like to overview the structure of the researched phenomenon. The research revealed multiple layers of the phenomenon.

1. Firstly, the research identified contradictions, or Janusian binaries, and benefits of social media application in university studies. Alongside the contradictions, there are benefits of social media application in university studies such as: the possibility to reach vast numbers of people almost instantly and wide access to information; information visualization and simulation (the possibility to use various sensory channels); and new forms of creativity and saving resources. The Janus-faced binaries are closely intertwined with the benefits of social media use and presuppose certain conclusions:

   (a) Wide access to information—the abundance of information requires skills or media "literacies" to deal with the constant flow of information.
   (b) The democratic scattered way of communicating on social media challenges the formal institutional hierarchical nature, at the same time changing the teacher role into more of a tutor and blending together formal and informal learning.
   (c) Privacy, seemingly guaranteed by the possibility of hiding behind the computer screen, seems to be challenged by social media use in university studies as the research participants experience it as an invasion of study and workplaces into their private spaces or as social media eliminating the defined limits between private and public.

2. Another layer of the phenomenon is phenomenological existentials: The first two, spatiality and temporality, are closely intertwined together, and the third one is relationality. This layer also provides certain conclusions:

   (a) The research participants experience spatiality as social media space used for various activities related to teaching and learning, which causes the changing perception or the redefinition of the working place.

(b) It is acutely felt by the teacher research participants that their time is sometimes perceived as stress or additional load, especially time devoted to mastering social media or answering the avalanche of student questions, which shows that social media use in university studies demands a new approach to teacher workload and its regulation. All these challenges pose a serious question about management of newly defined working places and understanding that teachers might feel simply overloaded at times. The issue should be addressed in an appropriate way, not just simply forcing teachers to look for the solutions or ways out on their own.

(c) Often the research participants experience time as lost time, as a loss or a distraction, which once more confirms the necessity of media "literacies."

3. Finally, at the deepest layer there are the two essential super-ordinate themes of pedagogical relationship and teacher creativity, which constitute the core of the phenomenon and also allow certain conclusions:

(a) The research proves that the pedagogical relationship and human creativity are at heart of technological "cyborgian" existence.

(b) There stands out the necessity of a sensitive democratic teacher–student pedagogical relationship characteristic of a cooperative atmosphere, acknowledging students' multitasking, facilitating student activities, flexibility, tolerance, and preference for informal communication. That is why it is necessary to use resources and time for equipping staff with the newest philosophical–pedagogical insights so that teachers acting as "trusted filters of information" or "knowledgeable others" could satisfy the need expressed by students.

(c) Although social media seems to redefine university study processes, as social media technology is a medium that redefines us, the pedagogical relation between teachers and students still stands out as being of core importance, which proves that the human factor remains essential even in a technology-wired world.

(d) There is an outcry for teacher availability and more communication in social media environments expressed by student research participants, while teacher research participants experience unlimited communication such as a threat to their privacy and invasion of working place and time into their private space.

(e) Teacher research participants experience their universities' institutional encouraging effect on their creativity, which helps teachers to acquire and use social media in university studies; however, some teacher research participants experience institutional influence as pressure or constraint.

(f) Creativity helps teacher research participants to find original solutions in unexpected situations while using social media in university studies, which proves that people make choices and decisions even if the question is changed along with the answer.

References

Adams, C. 2010. "Teachers Building Dwelling Thinking with Slideware". *The Indo-Pacific Journal of Phenomenology* 10 (1): 1–12.

Adams, C. 2012. "Technology as Teacher: Digital Media and Re-Schooling of Everyday Life". *Existential Analysis* 23 (2): 263–273.

Agichtain, E., C. Castillo, D. Donato, A. Gionis, and G. Mishene. 2008. "Finding high-quality content in social media." In *WSDM '08 Proceedings of the 2008 International Conference on Web Search and Data Mining*, 183–94. New York, NY. Accessed 5 June 2017. http://www.mathcs.emory.edu/~eugene/papers/wsdm2008quality.pdf.

Akakandelwa, A., Walubita, G. 2018. "Students' Social Media Use and its Perceived Impact on their Social Life: A Case Study of the University of Zambia." *The International Journal of Multi-Disciplinary Research.*

Alexander, B. 2006. "Web 2.0: A New Wave of Innovation for Teaching and Learning?" *EDUCAUSE Review* 41 (2): 32–44. Accessed 19 Jan 2018. https://er.educause.edu/articles/2006/1/web-20-a-new-wave-of-innovation-for-teaching-and-learning.

Amabile, T.M. 1983. *The Social Psychology of Creativity*. New York, NY: Springer-Verlag.

Amabile, T.M., R. Conti, H. Coon, J. Lazenby, and M. Herron. 1996. "Assessing the Work Environment for Creativity". *Academy of Management Journal* 39 (5): 1154–1184.

American Library Association. 1989. *Information Literacy Competency Standards for Higher Education*. Accessed 5 Apr 2018. http://www.ala.org/acrl/standards/informationliteracycompetency.

Anderson, L.W., and D. Krathwohl. 2001. *A Taxonomy for Learning, Teaching and Assessing: A Revision of Bloom's Taxonomy of Educational Objectives*. New York, NY: Longman.

Anderson, P. 2007. "What is Web 2.0? Ideas, Technologies and Implications for Education." *JISC Technology and Standards Watch*, 1–64.

Antulienė, A., V. Liubinienė, Z. Mažuolienė, I. Šneiderienė, V. Valiuškevičienė, and G. Valūnaitė Oleškevičienė. 2005. *A Survey of English Language Teaching in Lithuania: 2003–2004*. Vilnius: Britų taryba ir Lietuvos švietimo ir mokslo ministerija.

Arnold, M. 2003. "On the Phenomenology of Technology: The 'Janus-faces' of Mobile Phones". *Information and Organization* 13 (4): 231–256.

Ashton, J., and L. Newman. 2006. "An Unfinished Symphony: 21st Century Teacher Education Using Knowledge Creating Heutagogies". *British Journal of Educational Technology* 37 (6): 825–840.

Ausburn, L.J. 2004. "Environments: An American Perspective. Blended Learning Part 2". *Educational Media International* 41 (4): 327–337.

G. Valunaite Oleskeviciene and J. Sliogeriene, *Social Media Use in University Studies*, Numanities - Arts and Humanities in Progress 13, https://doi.org/10.1007/978-3-030-37727-4

Bach, S., P. Haynes, and J.L. Smith. 2007. *Online Learning and Teaching in Higher Education*. Oxford: Oxford University Press.

Balnchot, M. 1981. *The Gaze of Orpheus*. New York, NY: Station Hill Press.

Bandura, A. 1986. *Social Foundations of Thought and Action: A Social Cognitive Theory*. Englewood Cliffs, NJ: Prentice Hall.

Bandura, A. 2002. "Social Cognitive Theory in Cultural Context". *Applied Psychology: An International Review* 51 (2): 269–290.

Barak, M., A. Lipson, and S. Lerman. 2006. "Wireless Laptops as Means for Promoting Active Learning in Large Lecture Halls". *Journal of Research on Technology in Education* 38 (3): 245–263.

Baran, S.J., and D.K. Davis. 2009. *Mass Communication Theory*. Boston, MA: Wardsworth.

Barnett, R. 1990. *The Idea of Higher Education*. Buckingham: SRHE and Open University Press.

Bates, A. 2005. *Technology, e-Learning and Distance Education*. London: Routledge.

Bates, A.W. 1999. *The Impact of New Media on Academic Knowledge*. Accessed 1 May 2017. http://sdcc.vn/template/5298_knowledge.pdf.

Bauman, Z. 2011. *Vartojamas gyvenimas*. Vilnius: Apostrofa.

Beer, D., and R. Burrows. 2007. "Sociology and, of and in Web 2.0: Some Initial Considerations." *Sociological Research Online* 12 (5): 17. Accessed 26 June 2017. http://www.socresonline.org.uk/12/5/17.html.

Beetham, H., and R. Sharpe. 2007. *Rethinking Pedagogy for a Digital Age: Designing and Delivering E-Learning*. London: Routledge.

Benner, B.E. 1994. *Interpretive Phenomenology*. Thousand Oaks, CA: Sage Publications.

Bernoff, J., and C. Li. 2008. *Groundswell: Winning in a World Transformed by Social Technologies*. Boston: Harvard Business School Press.

Besley, T. (ed.). 2009. *Assessing the Quality of Educational Research in Higher Education: International Perspectives*. Rotterdam: Sense Publishers.

Biesta, G.J.J. 2013. *The Beautiful Risk of Education*. London: Paradigm Publishers.

Bloom's Revised Digital Taxonomy. 2011. Accessed 20 Apr 2018. http://www.celt.iastate.edu/teaching/effective-teaching-practices/revised-blooms-taxonomy.

Borgmann, A. 1987. *Technology and the Character of Contemporary Life: A Philosophical Inquiry*. Chicago, IL: University of Chicago Press.

Bouchanan, R., and A. Chapman. 2009. *The Sorry Story of the Digital Native*. Accessed 22 Apr 2017. https://researchbank.acu.edu.au/fea_pub/1837/.

Boyd, D., and N. Ellison. 2007. "Social Network Sites: Definition, History, and Scholarship." *Journal of Computer-Mediated Communication* 13 (1): 11. Accessed 12 June 2017. http://www.socialcapitalgateway.org/content/paper/boyd-d-m-ellison-n-b-2007-social-network-sites-definition-history-and-scholarship-jour.

Brown, T. 2006. "Beyond Constructivism. Navigationism in the Knowledge Era". *On the Horizon* 14 (3): 108–120.

Bryer, T.A., and B. Chen. 2012. "Investigating Instructional Strategies for Using Social Media in Formal and Informal Learning". *The International Review of Research in Open and Distance Learning* 13 (1): 87–104.

Burkšaitienė, N. 2017. "A lithuanian case of fostering creativity within academia: students' perceptions." In *Edu World 2016 7th International Conference. The European Proceedings of Social & Behavioural Sciences*. Accessed 3 July 3 2018. http://doi.org/10.15405/epsbs.2017.05.02.141.

Burkšaitienė, N., and M. Teresevičienė. 2004. "Innovative Learning and Assessment in Higher Education". *Tiltai* 19: 31–38.

Burkšaitienė, Nijolė, Jolita Šliogerienė. 2013. Assessment and recognition of non-formal and informal learning at university in Lithuania. In *Developing Sustainability : A Collection of Selected Papers Compiled by the Dorich House Group of Universities*, 61–76. İstambul : Bilgi University Press.

Burton, R.C. 1983. *The Higher Education System: Academic Organization in Cross-national Perspective*. Berkeley, CA: University of California Press.

Candy, Ph. 2004. *Linking Thinking: Self-directed learning in the Digital Age*. Accessed 13 Feb 2018. http://www.voced.edu.au/content/ngv%3A31516.

Carliner, S. 2005. "Course Management Systems Versus Learning Management Systems". *Learning Circuits*. http://www.learningcircuits.org/2005/nov2005/carliner.htm. Accessed 2 June 2018.

Carr, N. 2010. "The Web Shatters Focus, Rewires Brains." *Wired*. Accessed 21 May 2017. https://www.wired.com/2010/05/ff-nicholas-carr/.

Castells, M. 2000. *The Rise of the Network Society*. Oxford: Blackwell Publishers Ltd.

Castells, M. 2007. *Tūkstantmečio pabaiga*. Kaunas: Poligrafija ir informatika.

Chai, J., and K. Fan. 2018. "Constructing Creativity: Social Media and Creative Expression in Design Education". *EURASIA Journal of Mathematics, Science and Technology Education* 14 (1): 33–43.

Cheal, C., J. Coughlin, and S. Motore. 2012. *Transformation in Teaching: Social Media Strategies in Higher Education*. Santa Rosa, CA: Informing Science Press.

Chickering, A., and S.C. Ehrmann. 1996. "Implementing the Seven Principles: Technology as Lever." *AAHE Bulletin*, 3–6 Oct 1996. Accessed 9 Dec 2017. http://sphweb.bumc.bu.edu/otlt/teachingLibrary/Technology/seven_principles.pdf.

Christakis, N., and J. Fowler. 2009. *Connected*. New York, NY: Little Brown.

"Cisco Connected World Technology Report." 2011. Cisco press release, 21 Sept 2011. San Jose, CA. Accessed September 14, 2017. http://newsroom.cisco.com/press-release-content?articleId=474852.

Collis, B., and J. Moonen. 2001. *Flexible learning in a Digital World. Experiences and expectations*. London: Kogan Page Limited.

Collis, B., and M. Wende. (eds.). 2002. "Models of Technology and Change in Higher Education: An international Comparative Survey on the Current and Future Use of ICT in Higher Education by 2002." Center for Higher Education Policy Studies (CHEPS). Accessed 10 Sept 2017. http://doc.utwente.nl/44610/1/ictrapport.pdf.

Costley, C., and P. Armsby. 2007. "Work-Based Learning Assessed as a Field or a Mode of Study". *Assessment & Evaluation in Higher Education* 32 (1): 21–33.

Cowen, R. 2013. "Changing Principles and Goals of Universities: Questioning Trajectories". *Sisyphus Journal of Education* 1 (2): 38–53.

Creswell, J.W. 2007. *Research Design: Qualitative, Quantitative and Mixed Methods Approaches*, 2nd ed. Thousand Oaks, CA: Sage Publications.

Crook, C. 2008. "Theories of formal and informal learning in the world of web 2.0." In *Theorising the Benefits of New Technology for Youth*, ed. S. Livingstone. Oxford: Oxford University Press.

Cropley, A.J. 2008. *Creativity in Education and Learning*. London and New York: Routledge Falmer.

Csikszentmihalyi, M. 1996. *Creativity*. New York, NY: Harper Collins.

Dabbagh, N., and A. Kitsantas. 2012. "Personal Learning Environments, Social Media, and Self-Regulated Learning: A Natural Formula for Connecting Formal and Informal Learning". *The Internet and Higher Education* 15 (1): 3–8.

Daniel, J.S. 1998. *Mega Universities & Knowledge Media*. London: Kogan Page Limited.

Davenport, T.H., and J.C. Beck. 2001. *The Attention Economy: Understanding the New Currency of Business*. Boston: Harvard Business School Press.

De Angelis, K., N. Catenazzi, M. Graham, M. Klebl, V. Mazeikiene, G. Valunaite Oleskeviciene, K. Palfreyman, K, Reid, J. Sliogeriene, J. Tolonen, J. Van Zaalen. 2013. *ISTUS Report: Institutional Strategies for the Uptake of Social Media in Adult Education*. Vilnius: MRU ebooks.

Deleuze, G., and C. Parnet. 1987. *Dialogues*. London: The Athlone Press.

Denscombe, M. 2003. *The Good Research Guide*. Maidenhead, PA: Open University Press.

De Rossi. 2007. *Online Social Networking and Education: Study Reports an New Generations Social and Creative Interconnected Lifestyles*. November 9 2007. Accessed 4 June 2017 http://www.masternewmedia.org.

Derrida, J. 2000. *O grammatologii*. Maskva: Ad Marginem.

Dillenbourg, P. 2000. "Virtual Learning Environments" workshop. In *EUN Conference 2000: Learning in the New Millennium: Building New Education Strategies for Schools*. University of Geneva. Accessed 12 Apr 2017. http://tecfa.unige.ch/tecfa/publicat/dil-papers-2/Dil.7.5.18.pdf.

Dilthey, W. 1985. *Poetry and Experience*. Princeton University Press.

DiNucci, D. 1999. "Fragmented Future". *Print Magazine* 4: 32.

Donnelly, R. 2010. "Harmonizing Technology with Interaction in Blended Problem-Based Learning". *Computers & Education* 54 (2): 350–359.

Dougiamas, M., and P.C. Taylor. 2002. "Interpretive analysis of an internet-based course constructed using a new courseware tool called Moodle." In *2nd Conference of HERDSA (the Higher Education Research and Development Society of Australasia)*, 7–10. Accessed 8 June 2017. https://scholar.google.com/citations?user=R88V_rUAAAAJ&hl=en.

Downes, S. 2006. "Learning Networks and Connective Knowledge". Paper presented to IT Forum. Accessed 26 Sept 2017. https://www.downes.ca/files/books/Connective_Knowledge-19May2012.pdf.

Doyle, C.S. 1992. *Final Report to National Forum on Information Literacy*. New York: ERIC Clearinghouse on Information Resources.

Dreyfus, H.L. 1991. "Toward a phenomenology of ethical expertise". *Human Studies* 14: 229–250.

Dreyfus, H.L. 2008. *On the Internet (Thinking and Action)*. London: Routledge.

Dreyfus, H.L. 2012. "A History of First Step Fallacies". *Minds and Machines* 22: 87–99.

Drori, G.S., and J.W. Meyer. 2006. Scientization: Making a world safe for organizing. In *Transnational Governance: Institutional Dynamics of Regulation*, ed. Marie-Laure Djelic and Kerstin Sahlin-Andersson, 31–52. Cambridge: Cambridge University Press.

Dudeney, G., and N. Hockly. 2007. *How to teach English with Technology*. England: Pearson Education Limited.

Duffy, P.D., and A. Bruns. 2006. "The use of blogs, Wikis and RSS in education: A conversation of possibilities." In *Proceedings of the Online Learning and Teaching Conference 2006*, 31–38. Brisbane, Australia.

Duffy, T.M., and D.J. Cunningham. 1996. Constructivism: Implications for the design and delivery of instruction. In *Handbook of Research for Educational Communications and Technology*, ed. D.H. Jonassen. New York, NY: Macmillan Library Reference.

Duoblienė, L. 2011. *Ideologizuotos švietimo kaitos teritorijos*. Vilnius: Vilniaus universiteto leidykla.

Eijkman, H. 2008. "Web 2.0 as a Non-Foundational Network-Centric Learning Space". *Campus Wide Information Systems* 25 (2): 93–104.

Ellison, N., C. Steinfield, and C. Lampe. 2007. "The benefits of Facebook 'friends': Exploring the relationship between college students' use of online social networks and social capital." *Journal of Computer-Mediated Communication* 12 (3). Accessed 4 June 2017. https://www.academia.edu/2924152/The_benefits_of_Facebook_friends_Exploring_the_relationship_between_college_students_use_of_online_social_networks_and_social_capital.

Elo, S., and H. Kyngäs. 2007. "The Qualitative Content Analysis Process". *Journal of Advanced Nursing* 62 (1): 107–115.

Embree, L. 2001. "The Continuation of Phenomenology: A Fifth Period?". *Indo-Pacific Journal of Phenomenology* 1 (1): 1–7.

Epstein, D., R. Boden, R. Deem, F. Rizvi, and S. Wright (eds.). 2008. *Geographies of Knowledge, Geometries of Power: Higher Education in the 21st Century. World Yearbook of Education*. London: Kogan Page.

European Commission. 2004. *"Key Competences for Lifelong Learning. A European Reference Framework. Implementation of Education and Training 2010."* Work programme, Brussels: European Commission.

Evans, J. 2010. "Industry Collaboration, Scientific Sharing and the Dissemination of Knowledge." *Social Studies of Science* 40 (5): 757–91.

Evans, L. 2010. A phenomenological analysis of social networking. In *Humanity in Cybernetic Environments*, ed. D. Riha, 55–77. Oxford: Inter-disciplinary Press.

Ezzy, D. 2002. *Qualitative Analysis: Practice and Innovation*. Crows Nest, NSW: Allen & Unwin.

Feenberg, A. 1999. *Questioning Technology*. London: Routledge.

Fewkes, A.M., and M. McCabe. 2012. "Facebook: Learning Tool or Distraction?" *Journal of Digital Learning in Teacher Education* 28 (3): 92–98. Accessed 14 Apr 2018. http://search.ebscohost. com/login.aspx?direct=true&db=ehh&AN=72270772&site=ehost-live.

Focus on Higher Education in Europe 2010: The Impact of the Bologna Process. 2010. Education, Audiovisual and Culture Executive Agency, and Eurydice. Accessed 4 Feb 2018. http://eacea.ec. europa.eu/education/eurydice./documents/thematic_reports/122EN.pdf.

Foucault, M. 1977. *Discipline and Punish: The Birth of the Prison*. London: Peregrine Press.

Foucault, M. 1998. *Disciplinuoti ir bausti: kalėjimo gimimas*. Vilnius: Baltos lankos.

Franken, R.E. 1994. *Human Motivation*. California: Brooks/Cole Publishing Company.

Freire, P. 2007. *Pedagogy of the Oppressed*. New York, NY: Continuum.

Fried, C.B. 2008. "In-Class Laptop Use and Its Effects on Student Learning". *Computers and Education,* 50: 906–914.

Friesen, N., and S. Lowe. 2012. "Learning with Web 2.0: Social Technology and Discursive Psychology". *E-Learning and Digital Media* 9 (4): 377–387.

Friesen, N., and S. Lowe. 2012. "The questionable promise of social media". *International Journal of Computer Assisted Learning* 28 (3): 183–194.

Gadamer, H.-G. 1999. *Istorija, Menas, Kalba*. Vilnius: Baltos lankos.

Gadamer, H.-G. 1999. *Truth and Method*. London: Sheed and Ward.

Gardner, H. 1983. *Frames of Mind: The Theory of Multiple Intelligences*. New York: Basic Books.

Gedvilienė, G., and V. Vaičiūnienė. 2005. "Information Literacy Competency as a Premise for Successful Adult Education in the Civil Society." *International Perspectives in Adult Education*, 51, 69–82 (Adult Learning for Civil Society: The Institute for Cooperation of the German Adult Education Association).

Gill, M.J. 2014. "The Possibilities of Phenomenology for Organizational Research". *Organisational Research Methods* 17: 118–137.

Glastra, F.J., B.J. Hake, and P.E. Schedler. 2004. Lifelong Learning as Transitional Learning. *Adult Education Quarterly* 54 (4): 291–307.

Gouseti, A. 2010. Web 2.0 and Education: Not Just Another Case of Hype, Hope and Disappointment? *Learning, Media and Technology* 35 (3): 351–356.

Grant, I.H. 2001. "Postmodernism and Science and Technology." In *The Routledge Companion to Postmodernism*. London and New York: Stuart Sim.

Gredler, M.E. 2005. *Learning and instruction: Theory into practice*. Upper Saddle River, NJ: Pearson Merrill/Prentice Hall.

Gross, R., and A. Acquisti. 2005. "Information revelation and privacy in online social networks." In *ACM Workshop on Privacy in the Electronic Society (WPES)*, Alexandria, VA, Nov 7, 2005. Accessed 10 June 2017. http://www.heinz.cmu.edu/~acquisti/papers/privacy-facebook-gross-acquisti.pdf.

Hall, D., and R. Mansfield. 1995. "Relationships of Age and Seniority with Career Variables of Engineers and Scientists". *Journal of Applied Psychology* 60 (2): 201–210.

Hammond, N. 1994. "Learning Technology in Higher Education in Great Britain: Trends, Drivers and Strategies". *Social Science Computer Review* 12 (4): 585–609.

Hancock, A. 1998. "Contemporary information and communication technologies and education. Education for the twenty-first century." In *Issues and Prospects*. Paris: UNESCO Publishing.

Hargittai, E. 2007. "Whose Space? Differences Among Users and Non-Users of Social Network Sites." *Journal of Computer-Mediated Communication* 13 (1): 14–26. Accessed 6 July 2017. https://academic.oup.com/jcmc/article/13/1/276/4583068.

Hattie, J. 2008. *Visible Learning: A Synthesis of Over 800 Meta-Analyses Relating to Achievement*. London: Routledge.

Heemskerk, I., et al. 2005. "Inclusiveness and ICT in Education: A Focus on Gender, Ethnicity and Social Class". *Journal of Computer Assisted Learning* 21 (1): 1–16.

Heidegger, M. 1962. *Being and Time*. New York, NY: Harper and Row.

Heidegger, M. 1971. *Poetry, Language and Thought*, trans. A. Hofstadter. New York, NY: Harper and Row.

Heidegger, M. 1972. *The Basic Problems of Phenomenology*. Bloomington, IN: Indiana University Press.

Heidegger, M. 1977. *The Question Concerning Technology and Other Essays*, trans. and ed. W. Lovitt. New York, NY: Harper and Row.

Heidegger, M. 1985. *History of the Concept of Time*. Bloomington, IN: Indiana University Press.

Heidegger, M. 2000. *Introduction to Metaphysics*. New Haven, CT and London: Yale University Press.

Hembrooke, H., and G. Gay. 2003. "The Laptop and the Lecture: The Effects of Multitasking in Learning Environments". *Journal of Computing in Higher Education* 15 (1): 46–64.

Hennesy, S., et al. 2005. "Teacher Perspectives on Integrating ICT into Subject Teaching: Commitment, Constraints, Caution, and Change". *Journal of Curriculum Studies* 37 (2): 155–192.

Hewitt, A., and A. Forte. 2006. "Crossing Boundaries: Identity Management and Student/Faculty Relationships on the Facebook." Paper presented at the CSCW Conference, Banff, AB, Canada, 4–8 Nov 2006.

Hornecker, E. 2001. "Process and Structure—Dialectics Instead of Dichotomies." Position paper for E-CSCW Workshop on Structure and Process: The Interplay of Routine and Informed Action, Bonn, September 2001. Accessed 16 May 2017. http://www.ehornecker.de/Papers/TZI.pdf.

Hosein, A., R. Ramanau, and C. Jones. 2010. "Learning and Living Technologies: A Longitudinal Study of First-Year Students' Frequency and Competence in the Use of ICT." *Learning Media and Technology* 35 (4): 403–18 (Abingdon: Routledge, Taylor and Francis Group).

Houle, C. 1996. *The Design of Education*. San Francisco, CA: Jossey-Bass.

Hunt, S., L. Lippert, and S. Paynton. 1998. "Alternatives to Traditional Instruction: Using Games and Simulations to Increase Student Learning". *Communication Research Reports* 15 (1): 36–44.

Husserl, E. 1970. *The Idea of Phenemenology*. The Hague, The Netherlands: Nijhoff.

Ihde, D. 1990. *Technology and the Lifeworld*. Bloomington-Indianapolis, IN: Indiana University Press.

Ihde, D. 1993. *Post phenomenology: Essays in the Postmodern Context*. Evanston, IL: Northwestern University Press.

Ilf, I., and E. Petrov. 2011. *The Twelve Chairs*. Evanston: Northwestern university Press.

Illich, I. 1971. *Deschooling Society*. Reprinted in 1978. London: Marion Boyars.

Illich, I. 1996. "Philosophy … artifacts … friendship—and the history of the gaze." In *Philosophy of Technology: Proceedings of the American Catholic Philosophical Association*, ed. T.-A. Druart, 70, 61–82. Washington, DC: National Office of the American Catholic Philosophical Association, Catholic University of America.

International Bank for Reconstruction and Development. 2002. *Constructing Knowledge Societies: New Challenges for Tertiary Education*. Washington, DC. Accessed 10 Oct 2017. http://siteresources.worldbank.org/INTAFRREGTOPTEIA/Resources/Constructing_ Knowledge_Societies.pdf.

Irrgang, B. 2005. *Post-Human Humanity*. Stuttgart: Franz Steiner Verlag.

Jeffrey, B., and A. Craft. 2004. "Teaching Creatively and Teaching for Creativity: Distinctions and Relationships". *Educational Studies* 30 (1): 77–87.

Jeong, A.C., and S. Frassier. 2008. "How day of posting affects level of critical discourse in asynchronous discussions and computer-supported collaborative argumentation". *British Journal of Education Technology* 39 (5): 875–887.

Jespersen, P. 2001. "Redemptional Pedagogics." *Analytic Teaching* 20 (1). Accessed 20 May 2017. http://journaldatabase.info/journal/issn0890-5118.

Joinson, A., K. McKenna, T. Postmos, and U.D. Reips (eds.). 2007. *The Oxford Handbook of Internet Psychology*. Oxford: Oxford University Press.

Jones, C. 2004. "Networks and Learning: Communities, Practices and the Metaphor of Networks." *ALT—J: Research in Learning Technology* 12 (1): 81–93.

Jones, S., and S. Fox. 2009. *Generations Online in 2009*. Washington, DC: Pew Internet & American Life Project. Accessed 2 Apr 2017. http://www.pewinternet.org/files/old-media/Files/Reports/2009/PIP_Generations_2009.pdf.

Jovaiša, L. 1993. *Pedagogikos terminai*. Kaunas: Šviesa.

Jucevičienė, P., et al. 2010. *Universiteto edukacinė galia : atsakas XXI amžiaus iššūkiams*. Kaunas: Technologija.

Kaiser Family Foundation. 2010. Generation M^2 Media in the Lives of 8- to 18-Year-Olds. https://files.eric.ed.gov/fulltext/ED527859.pdf. Accessed 20 June 2018.

Kaplan, A.M., and M. Haenlein. 2010. "Users of the World, Unite! The Challenges and Opportunities of Social Media". *Business Horizons* 53 (1): 59–68.

Kellne, D. 2000. *New Technologies/New Literacies: Reconstructing Education for the New Millennium*. Graduate School of Education: University of Queensland.

Kennedy, G.E., et al. 2008. "First Year Students' Experiences with Technology: Are They Really Digital Natives?". *Australasian Journal of Educational Technology* 24 (1): 108–122.

Kenyon, C., and S. Hase. 2010. Andragogy and heutagogy in postgraduate work. In *Meeting the Challenges of Change in Postgraduate Education*, ed. T. Kerry, 108–197. London: Continuum Press.

Kietzmann, H.J., and K. Hermkens. 2011. "Social media? Get serious! Understanding the functional building blocks of social media". *Business Horizons* 54 (3): 241–251.

Kirkwood, A., and L. Price. 2014. "Technology-Enhanced Learning and Teaching in Higher Education: What Is 'Enhanced' and How Do We Know? A Critical Literature Review". *Learning, Media and Technology* 39 (1): 6–36.

Kirschner, P.A., and P. De Bruyckere. 2017. "The Myths of the Digital Native and the Multitasker". *Teaching and Teacher Education* 67: 135–142.

Kluitenberg, E. *Media without an Audience*. Accessed 3 July 2017. http://subsol.c3.hu/subsol_2/contributors0/kluitenbergtext.html.

Knowles, M.S., E.F. Holton, and R.A. Swanson. 2007. *Suaugęs besimokantysis: klasikinis požiūris į suaugusių švietimą*. Vilnius: Danielius.

Kop, R. 2010. "Networked Connectivity and Adult Learning: Social Media, the Knowledgeable Other and Distance Education." Doctoral thesis, Swansea University.

Kraujutytė, L. 2002. *Aukštojo mokslo demokratiškumo pagrindai*. Vilnius: LTU.

Kraushaar, J.M., and D.C. Novak. 2010. "Examining the Effects of Student Multitasking with Laptops During the Lecture". *Journal of Information Systems Education* 21 (2): 241–251.

Kvale, S. 1996. *InterViews: An Introduction to Qualitative Research Interviewing*. Thousand Oaks, CA: Sage Publications.

Lagemann, E.C. 2000. *An Elusive Science*. Chicago, IL: University of Chicago Press.

Lamb, B. 2007. "Dr. Mashup or, Why Educators Should Learn to Stop Worrying and Love the Remix". *Educause Review* 42 (4): 13–14.

Lao, T. 2009. Lao-Tzu's Tao Teaching. In *Port Townsend*, ed. B. Porter. WA: Copper Canyon Press.

Latour, B. 1993. *We Have Never Been Modern*. Cambridge, MA: Harvard University Press.

Laurillard, D. 1993. *Rethinking University Teaching—A Framework for the Effective use of Educational Technology*. London: Routledge.

Laurillard, D., 1999. "Using communications and information technology effectively." In *McKeachie's Teaching Tips: Strategies, Research and Theory for College and University Teachers*, ed. W.J. McKeachie and G. Gibbs. Boston, 183–200. MA: Houghton Miffin Co.

Laurillard, D. 2002. *Rethinking University Teaching: A Conversational Framework for the Effective Use of Learning Technologies*. London: RoutledgeFalmer.

Lave, J., and E. Wenger. 1991. *Situated Learning: Legitimate Peripheral Participation*. Cambridge: Cambridge University Press.

Lave, J., and E. Wenger. 2002. "Legitimate peripheral participation in communities of practice." In *Supporting Lifelong Learning: Volume 1, Perspectives on Learning*, ed. R. Harrison, et al., 111–26. London: Routledge Falmer.

Lee, M., and C. McLoughlin. 2010. *Web 2.0-Based e-Learning*. Hershey PA: Information Science Reference.

Leiblich, A. 1998. *Narrative Research: Reading, Analysis and Interpretation*. Thousand Oaks, CA: Sage Publications.

Lemke, J.L. 1998. Analysing verbal data: Principles, methods, and problems. In *International Handbook of Science Education*, ed. K. Tobin and B. Fraser, 1175–1189. London: Kluwer Academic Publishers.

Lenhart, A., K. Purcell, A. Smith, and K. Zickuhr. 2010. *Social Media and Mobile Internet Use among Teens and Young Adults*. Accessed 2 June 2017. http://www.pewinternet.org/files/old-media/Files/Reports/2010/PIP_Social_Media_and_Young_Adults_Report_Final_with_toplines.pdf.

LeNoue, M., T. Hall, and M.A. Eighmy. 2011. "Adult Education and the Social Media Revolution". *Adult Learning* 22 (2): 4–12.

Lievrouw, A., and S. Livingstone. 2002. *Handbook of New Media: Social Shaping and Consequences of ICTs*. London: Sage Publishing.

Lifelong Learning Strategy. 2003. *Mokymosi visą gyvenimą užtikrinimo strategija*. Vilnius: LR Švietimo ir mokslo ministerija.

Lindseth, A., and A. Norberg. 2004. "*A Phenomenological Hermeneutical Method for Researching*".

Liuveno. 2009. "LLN komunikatas Bolonijos procesas 2020- Europos aukštojo mokslo erdvė naujajame dešimtmetyje Communiqué of Bologna Ministers). Lived experience." *Scandinavian Journal of Caring Sciences* 18 (2): 145–153.

Livingstone, D.W. 2007. Conceptions of formal education and informal learning. In *Learning in Places*, ed. Z. Beckerman, 203–229. New York, NY: Peter Lang Publishing.

Lodico, M., D. Spaulding, and K. Voegtle. 2010. *Methods in Educational Research: From Theory to Practice*. San Francisco, CA: Wiley.

Loveless, A.M. 2002. "Literature Review in Creativity." *New Technologies and Learning. Report 4*. Accessed 16 Apr 2017. https://telearn.archives-ouvertes.fr/hal-00190439/document.

Lyotard, J.F. 1984. *The Postmodern Condition*. Manchester: Manchester University Press.

Macdonald, J. 2004. "Developing Competent E-Learners: The Role of Assessment". *Assessment and Evaluation in Higher education* 29 (2): 215–226.

MacKenzie, D., and J. Wajcman. 1999. *The Social Shaping of Technology*, 2nd ed. Buckingham: Open University Press.

Marquis J., 2012. *Pros and Cons of Social Media in Education*. http://www.onlineuniversities.com/blog/2012/02/pros-and-cons-of-social-media-in-education/. Accessed 4 Apr 2018.

Maslow, A.H. 1967. The creative attitude. In *Explorations in Creativity*, ed. R.L. Mooney and T.A. Rasik, 43–57. New York, NY: Harper and Row.

Mason, R. 2006. "Learning Technologies for Adult Continuing Education". *Studies in Continuing Education* 28 (2): 121–133.

Mason, R., and F. Rennie. 2008. *E-Learning and Social Networking Handbook. Resources for Higher Education*. Abingdon: Routledge.

Maund, B. 2003. *Perception*. Chesham: Acumen Publishing.

Mayer, A., and S.L. Puller. 2008. "The Old Boy (and Girl) Network: Social Network Formation on University Campuses". *Journal of Public Economics* 92 (1–2): 329–347.

Mayring, P. 2000. "Qualitative Content Analysis." *Forum: Qualitative Social Research* 1 (2). Accessed 4 Sept 2017. http://217.160.35.246/fqs-texte/2-00/2-00mayring-e.pdf.

Mayring, P. 2014. "*Qualitative Content Analysis. Theoretical Foundation, Basic Procedures and Software Solution*". *Social Science Open Access Repository*. Accessed 10 Sept 2017. http://www.psychopen.eu/fileadmin/user_upload/books/mayring/ssoar-2014-mayring-Qualitative_content_analysis_theoretical_foundation.pdf.

Mažeikienė, V., V. Vaičiūnienė, and G. Valūnaitė Oleškevičienė. 2013. *Social Media in Adult Education*. Vilnius: MRU.

Mazer, J., R. Murphy, and C. Simonds. 2007. "I'll See You on 'Facebook': The Effects of Computer-Mediated Teacher Self-Disclosure on Student Motivation, Affective Learning, and Classroom Climate". *Communication Education* 56 (1): 1–17.

McConnell, T.R., R.O. Berdahl, and M.A. Fay. 1973. *From Elite to Mass to Universal Higher Education: The British and American transformations*. Berkeley, CA: Center for Research and Development in Higher Education, University of California.

McCoy, B. 2013. "Digital Distractions in the Classroom: Student Classroom Use of Digital Devices for Non-Class Related Purposes". *Journal of Media Education* 4 (4): 5–15.

McLoughlin, C., and M.J.W. Lee. 2008. "Future Learning Landscapes: Transforming Pedagogy Through Social Software." *Innovate: Journal of Online Education* 4 (5). Accessed 9 June 2017. https://core.ac.uk/download/pdf/51073511.pdf.

McLuhan, M. 2003. *Kaip suprasti medijas. Žmogaus tęsiniai*. Vilnius: Baltos lankos.

Merleau-Ponty, M. 2002. *Phenomenology of Perception*, trans. C. Smith. New York, NY: Routledge.

Merriam-Webster Online Dictionary. n.d. Accessed 20 Apr 2017. http://www.merriam-webster.com.

Moeller S., A. Joseph, J. Lau, and T. Carbo. 2011. "Towards Media and Information Literacy Indicators". *Background Document of the Expert Meeting*, 4–6 Nov 2010, Bangkok, Thailand. Paris: UNESCO.

Moore, M.G. 1993. Three types of interaction. In *Distance Education: New Perspectives*, ed. K. Harry, M. John, and D. Keegan, 19–24. New York, NY: Routledge.

Morgan, N., G. Jones, and A. Hodges. 2012. *Social Media: The Complete Guide to Social Media from the Social Media Guys*. Creative Commons Attribution.

Moustakas, C. 1994. *Phenomenological Research Methods*. Thousand Oaks, CA: Sage Publications.

Munhall, P. 1989. "Philosophical Ponderings on Qualitative Research Methods in Nursing". *Nursing Science Quarterly* 2 (1): 20–28.

Murray, H.G. 1985. Classroom teaching behaviors related to college teaching effectiveness. In *Using Research to Improve Teaching*, ed. J.G. Donald and A.M. Sullivan, 21–34. San Francisco, CA: Jossey-Bass.

Nahl, D. 2001. "A Conceptual Framework for Explaining Information Behaviour". *Journal Studies in Media and Information literacy Education* 1 (2): 1–16.

Naidoo, R. 2008. "Higher education: A powerhouse for development in a neo-liberal age?" In *Geographies of Knowledge, Geometries of Power: Higher Education in the 21st Century. World Yearbook of Education*, ed. D. Epstein, R. Boden, R. Deem, F. Rizvi, and S. Wright, 248–65. London: Routledge.

National Research Council. 1999. *Being Fluent with Information Technology*. Washington, DC: National Academies Press.

National School Boards Association. 2007. "Online Social Networking and Education: Study Reports on New Generations Social and Creative Interconnected Lifestyles." *MasterNewMedia*, November 9, 2007. Accessed 6 Apr 2017. http://www.masternewmedia.org/learning_educational_technologies/social-networking/social-networking-in-education-survey-on-new-generations-social-creative-and-interconnected-lifestyles-NSBA-20071109.htm.

Oakeshott, M. 1989. The idea of university. In *The Voice of Liberal Learning/Michael Oakeshott on Education*, ed. T. Fuller. New Haven, CT and London: Yale University Press.

Oberg, H., and A. Bell. 2012. *Exploring Phenomenology for Researching Lived Experience in Technology Enhanced Learning*. Accessed 9 June 2017. https://www.academia.edu/1506652/Exploring_phenomenology_for_researching_lived_experience_in_Technology_Enhanced_Learning.

O'Reilly, T. 2005. *"Design Patterns and Business Models for the Next Generation of Software."* O'Reilly Media, Inc. September 30, 2005. Accessed 6 Mar 2018. http://oreilly.com/web2/archive/what-is-web-20.html.

Orlikowski, W.J. 2000. "Using technology and constituting structures: A practice lens for studying technology in organizations". *Organization Science* 11 (4): 404–428.

Oxford English Dictionary Online (OED). 2010. 3rd ed. Oxford: Oxford University Press.

Owen, M., L. Grant, S. Sayers, and K. Facer. 2006. "Social software and learning." In *Opening Education*, Futurelab, Bristol. Accessed 9 May 2013. https://www.researchgate.net/publication/32231458_Futurelab_Social_software_and_learning.

Paiva Franco, C. 2008. "How to Build Successful Engaging E-Learning Experiences". *Humanizing Language Teaching* 10 (2): 2–3.

Papacharissi, Z. 2010. *A Networked Self*. London: Routledge.

Papert, S. 1993. *The Children's Machine*. New York, NY: Basic Books.

Perniola, M. 2004. *The Sex Appeal in the Inorganic: Philosophies of Desire in the Modern World*. New York, NY: Bloomsbury.

Perry, R.P., and J. Smart (eds.). 1997. *Effective Teaching in Higher Education: Research and Practice*. New York, NY: Agathon.

Perry, R.P., and J. Smart (eds.). 2007. *The Scholarship of Teaching and Learning in Higher Education: An Evidence-Based Perspective*. Dordrecht: Springer.

Peters, H. 2005. "Contested Discourses: Assessing the Outcomes of Learning from Experience for the Award of Credit in Higher Education". *Assessment & Evaluation in Higher Education* 30 (3): 273–285.

Petherbridge, D. 2007. "Upgrading or Replacing Your Learning Management System: Implications for Student Support". *Online Journal of Distance Learning Administration*, 10 (1). http://www.westga.edu/~distance/ojdla/spring101/petherbridge101.htm. Accessed 18 May 2018.

Petkūnas V., and P. Jucevičienė. 2006. "Edukacinės paradigmos kaita IKT diegimo įtakoje: mokytojo ir mokinio vaidmenų įvertinimo kriterijai." *Socialiniai mokslai* 2 (52) (Kaunas, KTU).

Pineteh, E.A. 2011. "Using Virtual Interactions to Enhance the Teaching of Communication Skills to Information Technology Students". *British Journal of Educational Technology* 43 (1): 85–96.

Pittard, V. 2004. Evidence for E-learning Policy. *Technology, Pedagogy and Education* 13 (2): 181–193.

Pollock, D.L. 2013. *Designing and Teaching Online Courses*. Bainbridge State College, April 2013. Accessed 15 Mar 2018. http://fsweb.bainbridge.edu/QEP/Docs/DesigningandTeachingOnlineCourses.pdf.

Prensky, M. 2001. "Digital Natives, Digital Immigrants." *On the Horizon* 9 (5): 1–9. Accessed 10 Apr 2018. http://www.marcprensky.com/writing/Prensky%20-%20Digital%20Natives,%20Digital%20Immigrants%20-%20Part1.pdf.

Prensky, M. 2007. "How to Teach with Technology: Keeping both Teachers and Students Comfortable in an Era of Exponential Change". *Emerging Technologies for Learning, BECTA* 2 (4): 40–46.

Prensky, M. 2014. "VUCA: Variability, Uncertainty, Complexity, Ambiguity". *Educational Technology, LIV* 2: 64.

Rainie, L., A. Lenhart, and A. Smith. 2012. *The Tone of Life on Social Networking Sites*. A Project of the Pew Research Centre. February 9, 2012. Accessed 25 Nov 2017. http://pewinternet.org/Reports/2012/Social-networking-climate.aspx.

Ramirez, F.O. 2006. The rationalization of universities. In *Transnational Governance*, ed. M.L. Djelic and K. Sahlin-Andersson. Cambridge: Cambridge University Press.

Ramsden, P. 1992. *Learning to Teach in Higher Education*. London: Kogan Press.

Raskin, P. 2006. *The Great Transition Today: A Report from the Future*. Boston, MA: Tellus Institute.

Rheingold, H. 2010. "Attention, and Other 21st-Century Social Media Literacies". *Educause Review* 45 (5): 14–16.

Rickards, T.J. 1994. *Creativity from a Business School Perspective: Past Present and Future*. Norwood: Ablex.

Ricoeur, P. 2000. *Interpretacijos teorija: diskursas ir reikšmės perteklius*. Vilnius: Baltos lankos.

Robinson, K. 2001. *Out of Our Minds: Learning to be Creative*. Sussex: Capstone Publishing.

Rockman, I.F. and Associates. 2004. *Integrating Information Literacy into the Higher Education Curriculum: Practical Models for Transformation Information Literacy Curriculum*, 1st ed. San Francisco, CA: Jossey-Bass.

Rodrigues, P.J., and G. Lobato Miranda. 2013. "Personal Learning Environments: Conceptions and Practices". *RELATEC* 12 (1): 23–34.

Rogers, E.M. 1995. *Diffusion of Innovations*. New York, NY: Free Press.

Rothblatt, S. 1997. *The Modern University and Its Discontents: The Fate of Newman's Legacies in Britain and America*. Cambridge: Cambridge University Press.

Rouis, S., M. Limayem, and E. Salehi-Sangari. 2011. "Impact of Facebook Usage on Students' Academic Achievement: Role of Self-Regulation and Trust". *Electronic Journal of Research in Educational Psychology* 9 (3): 961–994.

Rouse, M. 2005. *TechTarget*. s.v. *"Definition ICT (information and communications technology, or technologies)"*. Accessed 26 June 2017. http://searchcio-midmarket.techtarget.com/definition/ICT.

Rudd, T., D. Sutch, and K. Facer. 2006. "Opening Education: Towards new learning networks." In *E-Learning and Social Networking Handbook*, ed. R. Mason and F. Rennie, 2008. London: Routledge.

Rudd, T., D. Sutch, and K. Facer. 2006. *Towards New Learning Networks*. Bristol: Futurelab. Accessed 12 June 2017. https://www.nfer.ac.uk/publications/FUTL56/FUTL56.pdf.

Saevi, T. 2011. "Lived Relationality as Fulcrum for Pedagogical-Ethical Practice". *Studies in Philosophy and Education* 30 (5): 455–461.

Saevi, T. 2012. "Seeing Pedagogically, Telling Phenomenologically: Addressing the Profound Complexity of Education". *Phenomenology and Practice* 6 (2): 50–64.

Sallnas, E.L., K. Rassmus-Grohn, and C. Sjostrom. 2000. "Supporting Presence in Collaborative Environments by Haptic Force Feedback". *ACM Transactions on Computer-Human Interaction* 7 (4): 461–476.

Salmi, J. 2009. *The Challenge of Establishing World-Class Universities*. Washington, DC: World Bank.

Samalavičius, A. 2010. *Universiteto idėja ir akademinė industrija*. Vilnius: Vilniaus pedagoginio universiteto leidykla.

Scardamalia, M., and C. Bereiter. 2006. Knowledge building: theory, pedagogy, and technology. In *Cambridge Handbook of the Learning Sciences*, ed. K. Sawyer, 97–118. New York: Cambridge University Press.

Scardamalia, M. 2002. Collective cognitive responsibility for the advancement of knowledge. In *Liberal Education in a Knowledge Society*, ed. B. Smith, 67–98. Chicago, IL: Open Court.

Scott, J., and P.J. Carrington. 2011. *The SAGE Handbook of Social Network Analysis*. London: Sage Publications.

Selwyn, N. 2007. *Web 2.0 Applications as Alternative Environments for Informal Learning—A Critical Review*. Paper for OECD-KERIS expert meeting. Accessed 3 Oct 2018. http://www.oecd.org/dataoecd/32/3/39458556.pdf.

Selwyn, N. 2009. "Face Working: Exploring Students' Education-Related Use of Facebook". *Learning, Media and Technology* 34 (2): 157–174.

Selwyn, N. 2012. "Social Media in Higher Education." *The Europa World of Learning 2012*. Accessed 15 July 2017. http://www.educationarena.com/pdf/sample/sample-essay-selwyn.pdf.

Selwyn, N., et al. 2006. *Adult Learning in the Digital Age*. London: Routledge Taylor and Francis Group.

Shirky, C. 2008. *Here Comes Everybody*. London: Allen Lane.

Siemens, G. 2004. "Connectivism: A Learning Theory for the Digital Age." *E-learnspace*. December 12, 2004. Accessed 10 June 2017. http://www.elearnspace.org/Articles/connectivism.htm.

Siemens, G. 2006. *Knowing Knowledge*. A Creative Commons licensed version. Accessed 5 Mar 2018. http://www.elearnspace.org/KnowingKnowledge_LowRes.pdf.

Siemens, G. 2011. "Orientation: Sense making and Way finding in Complex Distributed Online Information Environments." Doctoral thesis, University of Aberdeen.

Silverman, D. 2005. *Doing Qualitative Research: A Practical Handbook*. London: Sage Publications.

Šliogerienė, J., and G. Valūnaitė Oleškevičienė. 2014. "Confronting Social Media in Higher Education". *Socialinių mokslų studijos* 6 (2): 390–402.

Smith, J. A., P. Flower, and M. Larkin. 2009. *Interpretative Phonological Analysis. Theory, Method and Research*. London: Sage Publications.

Smith, T. 2009. "The Social Media Revolution". *International Journal of Market Research* 51 (4): 559–561.

Smith, H., and Smith, M. K. 2008. *The Art of Helping Others. Being Around, Being There, Being Wise*. London: Jessica Kingsley.

Skiba, D.J. 2013. "Bloom's Digital Taxonomy and Word Clouds". *Nursing Education Perspectives* 34 (4): 277–280.

Standish, P. 2008. "Preface". *Journal of Philosophy of Education* 42 (3–4): 349–353.

Stehr, N. 2005. *Knowledge Politics*. New York: Routledge.

Sternberg, R.J., and T.I. Lubart. 1999. *The Concept of Creativity: Prospects and Paradigms*. Cambridge: Cambridge University Press.

Stiegler, B. 1998. *Technics and Time, 1: The Fault of Epimetheus*. Stanford, CA: Stanford University Press.

Stiegler, B. 2010. *Taking Care of Youth and the Generations*, trans. S. Barker. Stanford, CA: Stanford University Press.

Straub, E. T. 2009. "Understanding Technology Adoption: Theory and Future Directions for Informal Learning." *Review of Educational Research* 79 (2): 625–49. Accessed 11 Nov 2017. http://rer.sagepub.com/content/79/2/625.

Strauss, A., and J. Corbin. 1990. *Basics of Qualitative Research*. Newbury Park: Sage Publications.

Subrahmanyam, K., and D. Šmahel. 2011. *Digital Youth*. Berlin: Springer.

Suoranta, J., and T. Vadén. 2010. *Wikiworld*. London: Pluto Press.

Sutherland, K., C. Davis, U. Terton, and I. Visser. 2018. "University Student Social Media Use and Its Influence on Offline Engagement in Higher Educational Communities". *Student Success* 9 (2): 13–24.

Tapscott, D. 1998. *Growing Up Digital: The Rise of the Net Generation*. New York, NY: McGraw-Hill.

Thomas, D., and J. Seely-Brown. 2011. *A New Culture of Learning*. Charleston, SC: Createspace.

Thomas, D.R. 2006. "A General Inductive Approach for Analyzing Qualitative Evaluation Data." *American Journal of Evaluation* 27 (2): 237–246. Accessed 11 Oct 2017. https://flexiblelearning.auckland.ac.nz/poplhlth701/8/files/general_inductive_approach.pdf.

Thomas, R.M. 2003. *Blending Qualitative and Quantitative Research Methods in Theses and Dissertations*. Thousand Oaks, CA: Corwin.

Thompson, I. 2005. *Heidegger on Ontotheology: Technology and the Politics of Education*. Cambridge: Cambridge University Press.

Thompson, J. 2007. "Is Education 1.0 Ready for Web 2.0 Students?". *Innovate* 3 (4): 1–8.

Tidikis, R. 2003. *Socialinių mokslų tyrimų metodologija*. Vilnius: Lietuvos teisės universitetas.

Ting, S.E. 2005. "The Impact of ICT on Learning: A Review of Research". *International Education Journal* 6 (5): 635–650.

Tinio, V.L. 2003. *ICT in Education*. New York. Accessed 10 Sept 2017. http://wikieducator.org/images/f/ff/Eprimer-edu_ICT_in_Education.pdf.

Tobin, K., and B. Fraser. 1998. *International Handbook of Science Education*, 1175–1189. London: Kluwer Academic Publishers.

Toole, T., P. Newrly, S. Pede, and L. Marcellin. 2010. "How to Promote Social Media Uptake in VET and Adult Training Systems in Europe—Practical Example of the European Project 'SVEA'." *eLearning Papers* 22. Accessed 11 Oct 2017. http://www.elearningpapers.eu/en/download/file/fid/19556.

Trowler, P. 2003. *Education Policy*. London and New York: Routledge.

Ulbrich, F., I. Jahnke, and P. Mårtensson. 2011. "Special Issue on Knowledge Development and the Net Generation." *International Journal of Sociotechnology and Knowledge Development* 14 (4): 241–52.

Underwood, J. 2004. Research into Information and Communication Technologies: Where Now? *Technology, Pedagogy and Education* 13 (2): 135–143.

Underwood, J., and G. Dillon. 2004. "Capturing Complexity Through Maturity Modelling". *Technology, Pedagogy and Education* 13 (2): 213–225.

UNESCO. 2005. "Towards Knowledge Societies. UNESCO World Report." Paris: UNESCO Publishing. Accessed 23 Apr 2018. http://unesdoc.unesco.org/images/0014/001418/141843e.pdf.

Van Manen, M. 1990. *Researching Lived Experience: Human Science for an Action Sensitive Pedagogy*. London, ON: The Althouse Press.

Van Manen, M. 1991. *The Tact of Teaching: The Meaning of Pedagogical Thoughtfulness*. London, ON: Althouse Press.

Van Manen, M. 1994. "On the Meaning of Pedagogy and its relation to Teaching". *Curriculum Inquiry* 4 (2): 135–170.

Van Manen, M. 1997. "From Meaning to Method". *Qualitative Health Research: An International, Interdisciplinary Journal* 12 (2): 264–280.

Van Manen, M. 2010. "The Pedagogy of Momus Technologies: Facebook, Privacy and Online Intimacy". *Qualitative Health Research* 20 (10): 1–10.

Van Manen, M. 2013. "The Call of Pedagogy as the Call of Contact". *Phenomenology and Practice* 6 (2): 8–34.

Van Manen, M. 2014. *Phenomenology of Practice*. California: Left Coast Press.

Veen, W., and B. Vrakking. 2006. *Homo Zappiens: Growing up in a Digital Age*. London: Network Continuum Education.

Venkatesh, V., M.G. Morris, G.B. Davis, and F.D. Davis. 2003. "User Acceptance of Information Technology: Toward a Unified View". *MIS Quarterly* 2 (3): 425–478.

Verbeek, P.P. 2008. "Cyborg Intentionality: Rethinking the Phenomenology of Human-Technology Relations". *Phenomenology and the Cognitive Sciences* 7 (3): 387–395.

Volungevičienė, A., and M. Teresevičienė. 2011. *Technologijomis grindžiamo mokymo(si) turinio kokybės vertinimas*. Kaunas: VDU.

Vovides, Y., S. Sanchex-Alonso, V. Mitropoulou, and G. Nickmans. 2007. "The Use of E-learning Course Management Systems to Support Learning Strategies and to Improve Self-regulated Learning". *Educational Research Review* 2 (1): 64–74.

Walberg, H.J., and W.E. Stariha. 1992. "Productive Human Capital: Learning, Creativity and Eminence". *Creativity Research Journal* 5 (4): 323–340.

Walker T, Siebert A. 1990. *Student Success: How to Succeed in College and Still Have Time for Your Friends*. Fort Worth: Holt Rinehart and Winston, pp 33–39.

Walters, P., and R. Kop. 2009. "Heidegger, Digital Technology and Post-Modern Education: From Being-In-Cyberspace to Meeting on Myspace". *Bulletin of Science, Technology and Society* 29 (4): 278–286.

Weber, F. 2006. *Theories of the Information Society*. Oxon: Routledge.

Weber, M. 1946. *Essays in Sociology*. New York, NY: Oxford University Press.

Weerakon, P. 2003. "Evaluation of online learning and students' perception of workload." Presented at The Higher Education Research and Development Society of Australia (HERDSA) 2003 Annual Conference. Accessed 8 Sept 2017. http://www.herdsa.org.au/publications/conference-proceedings/research-and-development-higher-education-learning-unknown-12.

Weller, M. 2007. *Virtual Learning Environments, Using, Choosing and Developing your VLE*. Abingdon: Routledge.

Wheeler, S. 2013. "The Meaning of Pedagogy." Accessed 3 Mar 2018. http://www.steve-wheeler.co.uk/2013/11/the-meaning-of-pedagogy.html.

Witzel, A., and H. Reiter. 2012. *The Problem-Centred Interview*. London, Thousand Oaks, New Delhi and Singapore: Sage Publications.

Young, J.R. 2006. "The Fight for Classroom Attention: Professor Versus Laptop". *Chronicle of Higher Education* 52 (39): A27–A29.

Young, W.P. 2007. *The Shack*. Los Angeles, CA: Windblown Media.

Zahavi, D. 2003. *Husserl's Phenomenology*. Stanford: Stanford University Press.

Zusman, A. 2005. "Challenges facing higher education in the twenty-first century." In *American Higher Education in the Twenty-First Century*, ed. P.G. Altbach, R.O. Berdahl, and P.J. Gumport, 2nd ed., 115–60. Baltimore: The Johns Hopkins University Press.

Zygmantas, J. 2007. "Changing Requirements for Language Teachers." *Socialiniai mokslai* 3 (57): 66–72 (Kaunas: KTU).

Index

G. Valunaite Oleskeviciene and J. Sliogeriene, *Social Media Use
in University Studies*, Numanities - Arts and Humanities in Progress 13,
https://doi.org/10.1007/978-3-030-37727-4